Molecular Genetics

METHODS IN HEMATOLOGY

Volume 20

EDITORIAL BOARD

I. Chanarin MD, FRCPath, *Chairman*
  Northwick Park Hospital and Clinical Research Centre, London, UK
Ernest Beutler MD
  Scripps Clinic and Research Foundation, La Jolla, California, USA
Elmer B. Brown MD
  Washington University School of Medicine, St Louis, Missouri, USA
Allen Jacobs MD, FRCPath
  Welsh National School of Medicine, Cardiff, UK

Also in the series
Volume 1: Iron, James D. Cook, *Guest Editor*
Volume 2: The Leukemic Cell, D. Catovsky, *Guest Editor*
Volume 3: Leucocyte Function, Martin J. Cline, *Guest Editor*
Volume 4: Quality Control, I. Cavill, *Guest Editor*
Volume 5: The Hemophilias, Arthur L. Bloom, *Guest Editor*
Volume 6: The Thalassemias, D. J. Weatherall, *Guest Editor*
Volume 7: Disorders of Thrombin Formation, Robert Colman, *Guest Editor*
Volume 8: Measurements of Platelet Function, Laurence A. Harker and Theodore S. Zimmerman, *Guest Editors*
Volume 9: The Immune Cytopenias, Robert McMillan, *Guest Editor*
Volume 10: The Cobalamins, Charles A. Hall, *Guest Editor*
Volume 11: Hematopoietic Stem Cells, David W. Golde, *Guest Editor*
Volume 12: Acquired Immune Hemolytic Anemias, Hugh Chaplin, *Guest Editor*
Volume 13: Monoclonal Antibodies, P. C. L. Beverley, *Guest Editor*
Volume 14: Radionuclides in Haematology, S. M. Lewis and R. J. Bayly, *Guest Editors*
Volume 15: Hemoglobinopathies, T. H. J. Huisman, *Guest Editor*
Volume 16: Red Cell Metabolism, Ernest Beutler, *Guest Editor*
Volume 17: Blood Transfusion, T. J. Greenwalt, *Guest Editor*
Volume 18: Venous Thrombosis and Pulmonary Embolism: Diagnostic Methods, Jack Hirsh, *Guest Editor*
Volume 19: Red Cell Membranes, Stephen B. Shohet and N. Mohandas, *Guest Editors*

Molecular Genetics

EDITED BY

Edward J. Benz, Jr. MD

Chief of Hematology Section,
Associate Chairman, Department of Internal Medicine,
Professor of Internal Medicine and Human Genetics,
Yale University School of Medicine

CHURCHILL LIVINGSTONE
EDINBURGH LONDON MELBOURNE AND NEW YORK 1989

CHURCHILL LIVINGSTONE
Medical Division of Longman Group UK Limited

Distributed in the United States of America by Churchill Livingstone Inc., 1560 Broadway, New York, N.Y. 10036, and by associated companies, branches and representatives throughout the world.

© Longman Group UK Limited 1989

All rights reserved. No part of this publication may be reproduced, stored in a retrieval system, or transmitted in any form or by any means, electronic, mechanical, photocopying, recording or otherwise, without either the prior written permission of the publishers (Churchill Livingstone, Robert Stevenson House, 1–3 Baxter's Place, Leith Walk, Edinburgh EH1 3AF), or a licence permitting restricted copying in the United Kingdom issued by the Copyright Licensing Agency Ltd, 33–34 Alfred Place, London, WC1E 7DP.

First edition 1989
ISBN 0-443-03852-X
ISSN 0264-4711

British Library Cataloguing in Publication Data

Molecular genetics.
  1. Man. Molecular genetics. Medical aspects
I. Benz, Edward J. II. Series
616′.042

Library of Congress Cataloging in Publication Data

Molecular genetics/edited by Edward J. Benz, Jr.
  p.   cm.
  Includes index.
  1. Molecular genetics — Technique.
  2. Hematology — Technique.
  I. Benz, Edward J.
  QH441.M65 1989   88-27412
  574.87′328 — dc19   CIP

Produced by Longman Singapore Publishers Pte Ltd
Printed in Singapore

Preface

Molecular genetics is a discipline dedicated to the understanding of the molecular basis for the regulated storage, expression and transmission to subsequent generations of genetic information. The term 'recombinant DNA technology' describes the methodological repertoire that has been devised by molecular biologists to permit direct examination of the nucleic acid (DNA and RNA) and protein molecules that represent the material transducers of this biological information management. The power of the molecular genetic approach lies in its ability to examine directly the basic kernel of genetic information flow in all living beings, namely the DNA or RNA sequence that forms each gene. This new discipline is revolutionizing the biological sciences because its strategies and methodology are universally and equally applicable to the study of virtually any living entity.

The era of molecular biology traces its origins to the landmark discoveries in the 1940s and 1950s that established DNA nucleotide sequences as the fundamental repository of genetic information throughout the biosphere. These advances led rapidly to the deciphering of the genetic code and the elucidation of the biochemical pathway of protein synthesis by which the code is expressed. What emerged was perhaps the most important insight in the history of biological research, a fundamental postulate that predicts that the magnificent diversity of species and behaviors encountered in nature is ultimately governed by the relatively simple rules of Watson-Crick base pairing. These rules constrain the chemical interactions allowed between two DNA or RNA nucleotide sequences, thus defining their ability to manage information. In order to realize the progress promised by this idea, methods were required to allow direct manipulation of DNA sequences. Recombinant DNA technology has provided this universal capability. Consequently, the molecular genetic paradigms can now be applied as readily to the study of hematologic phenomena as it is to the dissection of the bacterial and viral systems for which the approaches were first developed.

This volume attempts to introduce the *methods* of recombinant DNA technology to hematologists. The chapters that follow should be regarded as a 'core curriculum' that provides the essential techniques needed to introduce recombinant DNA technology into a typical research laboratory. Our goal is to facilitate in a practical way the efforts of investigators to get started. Toward this end, each author has been instructed to provide the rationale for a particular group of methods, an overview of the utility of the method, a step-by-step protocol of the actual method, and any 'tricks' or pitfalls of which one must be cognizant in order to develop the method successfully. In addition, each author has included brief discussions of selected alternate techniques. Therefore, each chapter should serve both as a core introduction and as a 'pointer' that allows the reader to access methodology appropriately customized to his or her needs.

This book does not offer a comprehensive introduction to the intellectual discipline of molecular genetics. For purposes of orientation, I have provided a brief survey (Chapter 1) of major tenets, but this merely skims the surface of the field. A number of excellent monographs that present molecular genetics in more depth are available. I have included some of these in the reading list at the end of Chapter 1. A Glossary defining the most commonly used terms and reagents of recombinant DNA technology has also been provided in order to facilitate familiarity with the necessary terminology.

The book is organized into several operationally-defined groups of subjects. Chapters 1–3 describe the basic steps needed to isolate DNA and RNA specimens from hematopoietic tissues; to separate and analyze the fragments by gel electrophoresis and restriction endonuclease mapping; and to label the DNA or RNA probes necessary for analysis of the samples by molecular hybridization techniques such as DNA or RNA blotting. For many investigators who wish to analyze the structure, function, or regulation of genes using probes already available because they have been cloned in other laboratories, Chapters 1–3 may be all that is necessary.

Chapters 4–6 outline the essential mechanics of gene cloning. Chapter 7 outlines a typical method for DNA sequencing. Since a wide variety of useful sequencing strategies is available, this chapter describes one approach *as an example* of the way to determine the primary structure (nucleotide sequence) of a cloned gene.

Several more specialized applications that have proved to be particularly useful in hematology are outlined in Chapters 8–11. These include: (1) an introduction to messenger RNA translation systems that allow one to examine synthesis and post-translational processing of polypeptide products; (2) the use of specialized DNA blotting strategies to examine the configuration of genes in chromatin with respect to their capacity for activity or inactivity; (3) the use of synthetic oligonucleotide probes; and, (4) *in situ* hybridization techniques for cytological and histochemical analysis of gene expression in individual cell types. Finally, Chapter 12 provides a brief introduction to the more complicated techniques of functional analysis of gene expression by transfection of genes into foreign host cells. This approach, called surrogate genetics, allows one to insert normal or altered genes into cells for functional analysis of their expression.

To the maximum extent possible, each chapter is free-standing. For example, an investigator wishing to learn how to isolate DNA and perform Southern gene blotting should be able to *locate* all of the essential methods by reading Chapter 1. However, Chapter 1 cannot *include* all of the related methods needed for successful Southern blotting. Each author has cross referenced other chapters where appropriate. For example, Southern gene blotting requires labeling of

DNA molecules for use as hybridization probes, a technique that is described in Chapter 3. It is our hope that cross referencing will allow an investigator seeking a specific method to turn directly to the appropriate chapter without a need to read through the entire book.

Each author has been encouraged to describe the methods actually in use in his or her laboratory as completely as possible, so there is some duplication of methods descriptions. Readers will note that methods devoted to the same goal are generally quite similar, but not identical. Individual laboratories exercise slight variations of standard procedures needed to extract DNA or RNA, label probes, etc. It is clearly not our intent to confuse the reader, but rather to assure him or her that it is possible to customize some of the detailed steps of these methods to fit individual circumstances.

The authors in this volume are active molecular biologists who have adapted the basic 'standard' methods to the systems under study in their laboratories. By comparing the steps that vary slightly from laboratory to laboratory, the reader will gain some perspective as to the 'tolerances' of these techniques. For example, the comprehensive description of methods for gene transfer into foreign host cells and analysis of gene expression therein (Chapter 12) describes many methods already introduced in other chapters. This redundancy is intentional. By following the methodology in Chapter 12, one can be assured that those exact methods have been used successfully for the precise purpose of constructing and analyzing gene transfer systems. In the event that difficulties are encountered, one can adopt one of the variations presented elsewhere in the text.

Each author has included a reading list and references that should permit readers to explore particular methods in more detail, to seek alternative strategies, or to learn more about the theoretical basis for the technology. In particular, the comprehensive *Manual of Molecular Cloning*, by Drs. T. Maniatis and J. Sambrook, is an excellent reference source. The goal of our volume is to provide a more compact and manageable compendium for starting molecular genetics in one's own laboratory. Most readers will find that the *Manual of Molecular Cloning* provides a useful companion reference source in much the same way that the city telephone directory complements and supplements the utility of one's personal address book.

As editor of this volume, I would like to extend my personal thanks to the authors who collectively form one of the most productive groups of young molecular biologists in academic hematology. They have taken time from a very busy research commitment to share methods that they have honed in their laboratories. We in turn are deeply indebted to Ms. Phoebe Barron, not only for excellent organization of the volume and preparation of major portions of the manuscript, but also for significant assistance in the preparation of the Glossary and Index. Dr. Ernest Beutler graciously provided preliminary review of both the organization of the volume and early drafts of these chapters. Ms. Susan Baserga, author of Chapter 11 Part One, and an MD/PhD graduate student in my laboratory, provided equally valuable opinions by reading drafts of the chapters along the way.

A perquisite of editorship is the privilege to use a preface such as this as a forum to thank those whose efforts and inspiration allowed one to develop an academic career. In that spirit, it is particularly pleasurable to extend public thanks to Drs. Arthur Pardee and George Russell, who introduced me to the magic of molecular genetics; to Drs. Bernard Forget, Arthur Nienhuis and David Nathan, whose support and inspiration as role models have nurtured my development as a hematologist; my wife Bobbie and my parents for unwavering support and understanding; and to the many colleagues, students, and associates who have made my 15 years in hematology research a truly fulfilling and exciting adventure.

New Haven, Connecticut E. J. B.
1989

Contributors

Arthur Bank MD
Professor of Medicine, Professor of Genetics and Development; Director, Division of Hematology, Columbia University, New York, USA

Susan J. Baserga MD PhD
Post-Doctoral Fellow, Department of Molecular Biophysic and Biochemistry, Yale University School of Medicine, New Haven, Connecticut, USA

Edward J. Benz, Jr. MD
Chief, Hematology Section; Associate Chairman, Department of Internal Medicine; Professor of Internal Medicine and Human Genetics, Yale University School of Medicine, New Haven, Connecticut, USA

Francis S. Collins MD PhD
Assistant Professor of Internal Medicine and Human Genetics, The University of Michigan School of Medicine, Ann Arbor, Michigan, USA

Maryann Donovan-Peluso PhD
Postdoctoral Research Scientist, Department of Genetics and Development, Columbia University, New York, USA

James Downing MD
Assistant Member, Department of Pathology, St Judes Childrens Research Hospital, Memphis, Tennessee, USA

Gordon Ginder BS MD
Associate Professor of Internal Medicine and the Genetics PhD Program; Associate Director of Hematology-Oncology, University of Iowa, Iowa, USA

Katherine A. High MD
Assistant Professor of Medicine and Pathology, University of North Carolina School of Medicine, Chapel Hill, North Carolina, USA

Jeffrey A. Kant MD PhD
Associate Professor, Pathology and Laboratory Medicine, University of Pennsylvania; Director, Hematology and Molecular Diagnosis Laboratory, Hospital of the University of Pennsylvania, Philadelphia, Pennsylvania, USA

David Leibowitz BA MD
Assistant Professor of Medicine, Columbia University; Assistant Attending Physician, Presbyterian Hospital, New York, USA

Stephen A. Liebhaber MD
Associate Investigator, Howard Hughes Medical Institute; Associate Professor of Human Genetics and Medicine, University of Pennsylvania, Pennsylvania, USA

Karen J. Lomax MD
Senior Staff Fellow, Bacterial Diseases Section, Laboratory of Clinical Investigation, National Institute of Allergy and Infectious Diseases, National Institutes of Health, Bethesda, Maryland, USA

Robert W. Mercer PhD
Assistant Professor of Cell Biology and Physiology, Washington University School of Medicine, St Louis, Missouri, USA

Mortimer Poncz MD
Assistant Professor of Paediatrics, The Children's Hospital of Philadelphia, The University of Pennsylvania School of Medicine, Pennsylvania, USA

Edward V. Prochownik MD PhD
Associate Professor of Pediatrics, Section of Hematology/Oncology, C. S. Mott Children's Hospital; Committee of Cellular and Molecular Biology, The University of Michigan School of Medicine, Ann Arbor, Michigan, USA

Thomas A. Rado MD PhD
Assistant Professor of Medicine and Microbiology, Division of Hematology and Oncology; Associate Scientist, The Comprehensive Cancer Center, University of Alabama at Birmingham, Alabama, USA

Lee Ratner MD PhD
Assistant Professor of Medicine and Microbiology, Washington University; Staff Physician, Barnes Hospital, Washington, D.C., USA

Jay W. Schneider MD PhD Candidate
Department of Human Genetics, Yale University School of Medicine, New Haven, Connecticut, USA

Darrel W. Stafford PhD
Professor of Biology, University of North Carolina, Chapel Hill, North Carolina, USA

Manuel F. Utset MD PhD Candidate
Department of Human Genetics, Yale University School of Medicine, New Haven, Connecticut, USA

Jerry Ware PhD
Division of Experimental Hemostasis, Department of Basic and Clinical Research, Scripps Clinic and Research Foundation, La Jolla, California, USA

Katherine Young DPhil
Senior Research Scientist, United Biomedical, Inc., Lake Success, New York, USA

Contents

1. Introduction to molecular genetics and recombinant DNA technology 1
   E. J. Benz
2. Isolation of genomic DNA, restriction endonuclease mapping and Southern gene blotting 21
   K. A. High   J. L. Ware   D. W. Stafford
3. RNA isolation and processing 37
   D. Leibowitz   K. Young
4. Hybridization probes in molecular hematology 51
   T. A. Rado   L. Ratner   J. M. Downing
5. cDNA and genomic cloning 72
   E. V. Prochownik   F. S. Collins
6. Molecular cloning of genes: use of hybridization probe screening methods 88
   J. A. Kant
7. Molecular cloning of genes: use of antibody screening methods 100
   R. W. Mercer
8. Analysis of DNA methylation and DNase I sensitivity in gene-specific regions of chromatin 111
   G. Ginder
9. DNA analysis by M13-dideoxy sequencing 124
   M. Poncz
10. Detection of mRNA in frozen tissue sections by *in situ* hybridization with RNA probes 139
    J. W. Schneider   M. F. Utset
11. Oligonucleotide design and use; *in vitro* mutagenesis 148
    PART ONE Oligonucleotide-directed site-specific mutagenesis 148
    S. J. Baserga
    PART TWO Oligonucleotide design and use 158
    K. J. Lomax

12. Use of *in vitro* translation techniques to study gene expression 169
    S. A. Liebhaber

13. Transfection of DNA into mammalian cells 181
    M. Donovan-Peluso   A. Bank

Glossary 195

Index 205

1
Introduction to molecular genetics and recombinant DNA technology

E. J. Benz

A universal characteristic of living beings is their ability to store, express, and transmit information. The basic repositories of biological information flow are called genes; genes consist of molecules of DNA or, in retroviruses, of RNA. Molecular geneticists attempt to understand the molecular basis for the flow and regulation of genetic information by using recombinant DNA methods to isolate and analyze genes. In this chapter I introduce the basic concepts and experimental strategies that serve as the theoretical basis for the methods outlined in subsequent chapters.

For nearly two decades after the elucidation of the crystallographic structure of DNA, molecular geneticists were largely confined to the study of simple microorganisms. Complex genomes of 'higher' organisms seemed to be beyond the capability of the limited repertoire of biochemical methods then available for manipulating DNA molecules. However, in the early 1970s a new approach, recombinant DNA technology, arose by the coalescence of advances in enzymology, nucleic acid biochemistry, and microbial genetics. This technology allows one to 'cut and paste' DNA molecules from diverse sources together to form novel combinations of genes, and then to introduce the DNA into new host cells, where the novel sequences can be propagated and expressed. Taken together, these capabilities can be exploited to achieve physical isolation or 'cloning' of individual genes, even if they originally represent only one part per million or less of a complex genome. Recombinant DNA technology thus offers a quantum increase in our capability to examine genes directly as individual physical entities.

The real power of molecular genetic approaches resides in the universality with which these relatively simple methods can be productively applied. Thus, the basic approach one might use to study normal gene regulation during hematopoiesis is equally applicable to the analysis of abnormal gene expression occurring in inherited diseases of coagulation, hemoglobin production, or blood cell proliferation. The methods outlined in this book are in widespread use not only in hematology laboratories, but in those of microbiologists, botanists and entomologists. It follows that the principles and methods described in this Introduction are useful to all areas of hematology.

This chapter begins with a brief overview of the principles of molecular genetics. These principles provide the framework for understanding how the structure of DNA molecules can control the anatomy, physiology, and behavior of organisms. The basic elements of the recombinant DNA approach will then be discussed, followed by a summary of a few practical considerations for setting up a molecular biology section in one's laboratory.

PRINCIPLES OF MOLECULAR GENETICS

The fundamental kernel of biological information storage is the gene. Genes were originally characterized as mathematical units of inheritance but are now known to consist of molecules of deoxyribonucleic acid (DNA). DNA molecules are extremely long unbranched polymers consisting of nucleotide subunits. Each nucleotide in turn contains a sugar moiety called deoxyribose, a phosphate group attached to the $5'$ carbon position, and a purine or pyrimidine base attached to the $1'$ position (Fig. 1.1). The linkages in the chain are formed by phosphodiester bonds between the $5'$ position of each sugar residue and the $3'$ position of the adjacent residue in the chain (see Fig. 1.1). The sugar phosphate links form the backbone of the polymer, while the nitrogenous purine or pyrimidine bases project perpendicular to the chain.

The four nitrogenous bases in DNA are the purines, adenosine and guanosine, and the pyrimidines, thymine and cytosine. The basic chemical configuration of ribonucleic acid (RNA), is quite similar, except that the sugar is ribose (having a hydroxyl group attached to the $2'$ carbon, rather than the hydrogen found in deoxyribose), and that the pyrimidine base uracil is used in place of thymine. The nitrogenous bases are commonly referred to by a shorthand notation: the letters A, C, T, G, and U are used to refer to adenosine, cytosine, thymine, guanosine, and uracil, respectively.

The ends of DNA and RNA strands are chemically distinct. Because of the $3' \rightarrow 5'$ phosphodiester bond linkage that ties adjacent bases together (Fig. 1.2), one end of the strand ($3'$ end) will have an unlinked (free) $3'$ sugar position, and the other (the $5'$ end), a free $5'$ position. There is thus a 'polarity' to the sequence of bases in a DNA strand. The same sequence read in a $3' \rightarrow 5'$ direction carries a different meaning than if read in a $5' \rightarrow 3'$ direction. Cellular enzymes can distinguish one end of a nucleic acid from the other; most enzymes that can 'read' along the DNA sequence tend to do so only in one direction ($3' \rightarrow 5'$ or $5' \rightarrow 3'$, but not both).

The ability of DNA molecules to encode unique parcels of information resides in the sequence of nitrogenous bases arrayed along the polymer chain. Under the physiologic conditions extant within living cells, DNA is thermodynamically most stable when two strands coil around each other to form a double stranded helix. The strands are aligned in an 'antiparallel' direction, that is, the strands have opposite $3' \rightarrow 5'$ polarity (see Fig. 1.2). This structure is stable only when the sugar phosphate backbones are arrayed on the outside of the helix with the nitrogenous bases stacked in the center (Fig. 1.2).

The two strands are held together (stabilized) by hydrogen bonds between the nitrogenous bases on one strand and the nitrogenous bases on the opposite

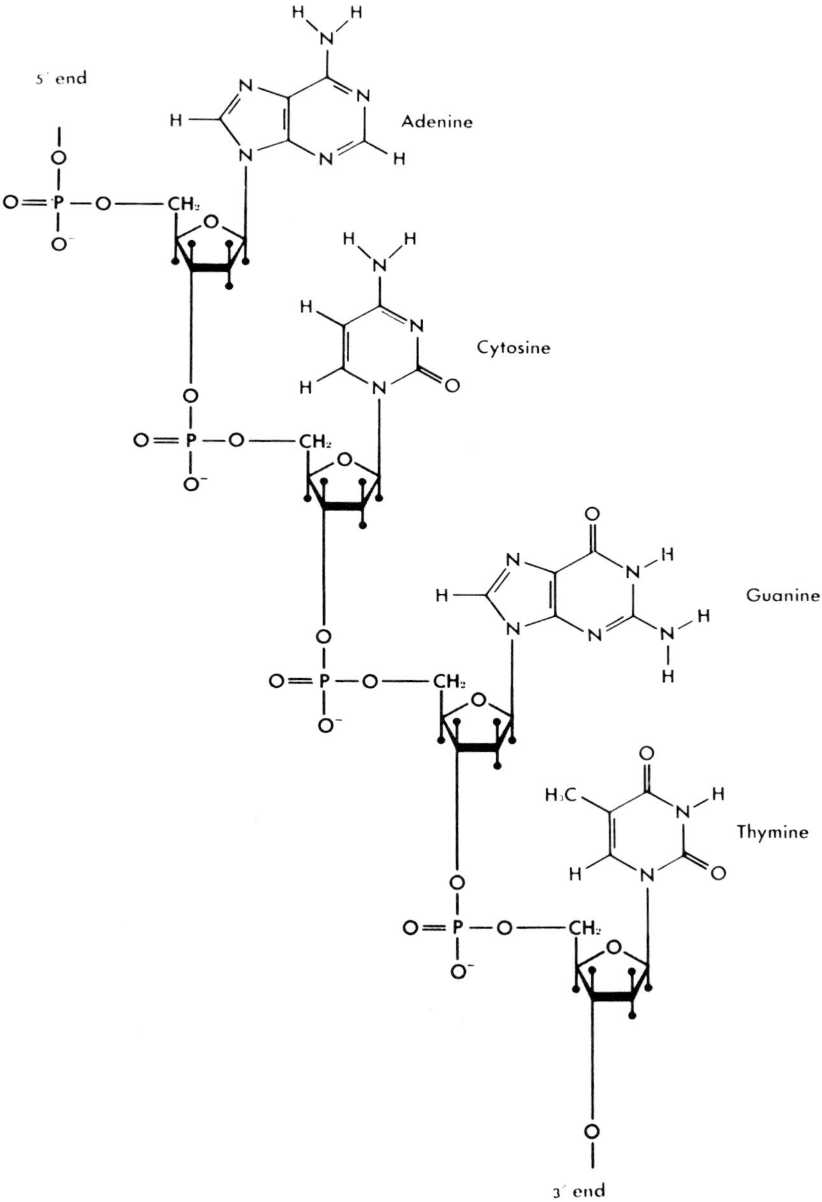

Fig. 1.1 The structure of DNA. The figure shows a portion of a single strand of DNA, demonstrating the projection of the purine and pyrimidine bases from a sugar-phosphate backbone. Note the 5′ → 3′ polarity from the top to bottom of the figure. See text for details.

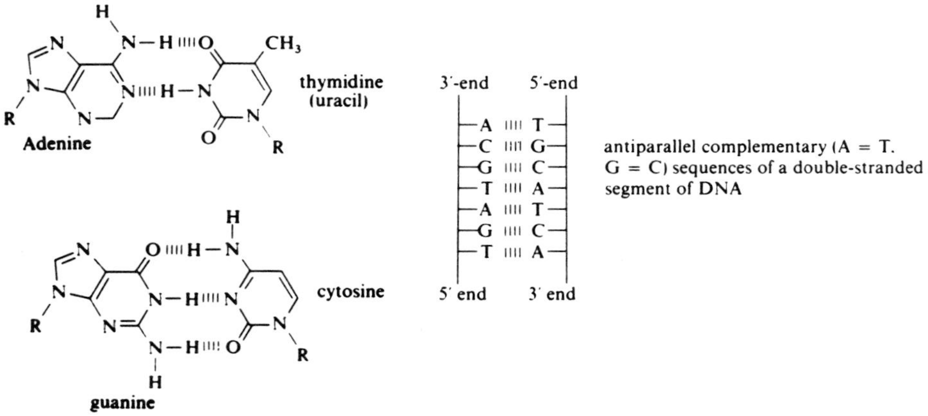

Fig. 1.2 Formation of base pairs according to the 'Watson-Crick' rules. Note the antiparallel (5' → 3') alignment of the two strands shown on the right side of the figure. See text for details.

strand. The stereochemical constraints of these hydrogen bond interactions dictate that hydrogen bonds can form between the two strands only if adenine on one strand pairs with thymine on the opposite strand, and guanine with cytosine. In other words, an adenine occurring at a certain position along a DNA strand can only bind effectively to a DNA strand having a thymine at the analogous position along the opposite strand. These 'Watson-Crick' rules of base pairing are usually expressed by saying that only A-T and G-C base pairs can form. Two strands joined together in compliance with these rules are said to have 'complementary' base sequences.

The implications of these thermodynamic rules are apparent: the sequence of bases along one DNA strand immediately dictates the sequence of bases that must be present along the opposite or complementary strand in the double helix. For example, whenever an A occurs along one strand a T must be present on the opposite strand; a G must always be paired with C, a T with an A, and a C with a G. This hydrogen bonding specificity confers on DNA strands their informational capacity. (Note: in RNA, U-A base pairs replace T-A base pairs.)

The rules of Watson-Crick base pairing apply to DNA-RNA, and RNA-RNA double stranded molecules, as well as DNA-DNA molecules. Enzymes that replicate or polymerize DNA and RNA molecules also obey the base pairing rules. They utilize an existing strand of DNA or RNA as the 'template'. The new (daughter) strand is then transcribed, or copied, by reading progressively along the base sequence of the template strand and, at each position, adding to the growing strand only that nitrogenous base that is 'complementary' to the base in the template by the Watson-Crick rules. Thus, a DNA strand having the base sequence 5'-AATGGC-3' could only be copied by DNA polymerase into a daughter strand having the sequence, 3'-TTACCG-5'. Note that the sequence of the template strand immediately provides all the information needed to predict the nucleotide sequence of the 'complementary' daughter strand.

Consider for a moment a double stranded DNA molecule that is separated into the two daughter strands. If each strand is then used as a template to synthesize a new daughter strand, what results is the creation of two double stranded daughter DNA molecules, each identical to the original parent molecule. This 'semi-conservative' replication process is exactly what occurs during mitosis and meiosis as cell division proceeds. In this manner the rules of Watson-Crick base pairing provide for the ability of DNA molecules to transmit faithful copies of themselves to subsequent generations.

The information stored in the DNA base sequence achieves its impact on the structure, function, and behavior of organisms by governing the structures and amounts of protein synthesized in the cells. The primary structure (amino acid sequence) of each protein determines its three dimensional conformation, and, therefore, its structural and functional properties (e.g. enzymatic activity, ability to interact with other molecules, stability, etc.). These proteins are the enzymes and structural elements that control cell structure and metabolism. To a first approximation, then, the structure and behavior of an organism is governed by the aggregate effects of the array of proteins produced. Genes determine the structures of proteins that are synthesized, the timing of their production during development or differentiation, and the amounts produced in different cells or tissues. In this manner DNA sequences control the properties of the organism. The process by which DNA achieves its control of cells is called gene expression.

A schema outlining the basic elements of gene expression is shown in Figure 1.3. The nucleotide base sequence in DNA is first copied into an RNA molecule, called messenger RNA, by mRNA polymerase. The mRNA has a base sequence complementary to the DNA coding strand. (Messenger RNA polymerase, like DNA polymerases, must obey the rules of Watson-Crick base pairing during transcription of the DNA gene into an mRNA strand.) Genes in all species except certain microorganisms consist of tandem arrays of sequences encoding messenger RNA (exons) that alternate with sequences present in the initial mRNA transcript, or precursor, but absent from the mature mRNA (introns). The entire gene is transcribed into the large precursor, which is further processed ('spliced') in the nucleus so that the RNA regions complementary to the introns are excised and discarded. The mRNA is then exported to the cytoplasm, where it is decoded and translated into the amino acid sequence of the protein by association with a biochemically complex group of ribonucleoprotein structures called ribosomes.

Ribosomes read the mRNA sequence in a 'ticker tape' fashion *three bases at a time*, inserting the appropriate amino acid encoded by each three base code word, or codon, into the appropriate position of the growing protein chain. This process is called messenger RNA translation. Thus, DNA regulates the properties of organisms by expression in the form of protein synthesis. Genetic information flows in the direction DNA → mRNA → protein. This polarity of information flow has been called the 'central dogma' of molecular biology.

The 'Rosetta stone' used by cells to know which amino acids are encoded by each DNA codon is called the genetic code (Table 1.1). The genetic code was

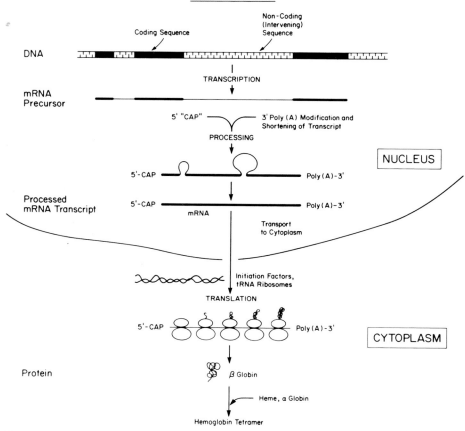

Fig. 1.3 The pathway of gene expression. Transcription and translation of the β-globin gene are shown as an example, but the basic steps shown here are typical of expression of most genes. See text for details.

deciphered by a series of elegant experiments conducted in several laboratories in the 1950s and 1960s. Each amino acid is encoded by a sequence of three successive bases. Recall that a sequence read in the $5' \rightarrow 3'$ direction has a different biological meaning than a sequence read in the $3' \rightarrow 5'$ direction. Given this polarity, (and an alphabet of the four code letters, A, C, U, and G) there are 4^3 or 64 possible three base codons.

There are 21 naturally occurring amino acids found in proteins, so that there are more codons available than there are amino acids to be encoded. As noted in Table 1.1, this redundancy or degeneracy of the genetic code results in the fact that some amino acids are encoded by more than one codon. For example, there are 6 possible codons that specify incorporation of serine at a specific position in the amino acid chain, and 4 codons for valine, but only one for methionine or tryptophan. However, in no case does a single codon encode more

Table 1.1 The genetic code. Messenger RNA codons for the amino acids.

Alanine	Arginine	Asparagine	Aspartic Acid	Cysteine
GCU	CGU	AAU	GAU	UGU
GCC	CGC	AAG	GAC	UGC
GCA	CGA			
GCG	CGG			
	AGA			
	AGG			
Glutamic acid	Glutamine	Glycine	Histidine	Isoleucine
GAA	CAA	GGU	CAU	AUU
GAG	CAG	GGC	CAC	AUC
		GGA		AUA
		GGG		
Leucine	Lysine	Methionine	Phenylalanine	Proline*
UUA	AAA	AUG**	UUU	CCU
UUG	AAG		UUC	CCC
CUU				CCA
CUC				CCG
CUA				
CUG				
Serine	Threonine	Tryptophan	Tyrosine	Valine
UCU	ACU	UGG	UAU	GUU
UCC	ACC		UAC	GUC
UCA	ACA			GUA
UCG	ACG			GUG
AGU				
AGC				
		Chain termination codons		
		UAA		
		UAG		
		UGA		

** AUG is also used as the chain initiation codon.
* Hydroxyproline, the 21st amino acid, is generated by post-translational modification of proline.

than one amino acid. Codons thus predict unambiguously the amino acid sequence they encode but one cannot easily read 'backward' from the amino acid sequence to decipher the encoding DNA sequence.

Some specialized codons serve as punctuation points in the genetic 'sentence'. The initiator codon, AUG, codes for methionine but also serves to signal the position at which to start protein synthesis. Three codons, UAG, UAA, and UGA serve as terminators marking the end of translation. These codons do not specify incorporation of amino acid. Rather, they inform the ribosomal apparatus that the amino acid chain has been completed and that dissociation of the ribosomal subunits from the mRNA should occur.

The adaptor molecules which mediate individual decoding events during mRNA translation are called transfer RNAs. These are small RNA species, approximately 40 nucleotides long. When each tRNA is bound into a ribosome it is configured in such a way that it exposes a 3 base segment, called the 'anticodon'. These 3 bases abut against the 3 base codon exposed on the mRNA that is also bound to the ribosome. Only transfer RNAs having a 3 base 'anticodon' complementary in sequence to the exposed 3 base codon in the mRNA will form a stable interaction among the mRNA, the ribosome, and the tRNA molecule. Within each tRNA is a separate region that is adapted for binding to an amino acid. The enzymes that catalyze the binding of the amino acid are constrained so that each tRNA species can only bind to a single amino acid. For example, tRNA molecules containing the anticodon 3'-TAC-5', which is complementary to a 5'-AUG-3' (methionine) codon in mRNA, can only be bound to or charged with methionine; tRNA containing the anticodon 3'-AAA-5' can only be charged with phenylalanine, etc.

The properties of tRNA and aminoacyl tRNA synthetase enzymes provide the specificity for translation of the genetic code. Ribosomes provide the reading apparatus by which tRNA anticodons and mRNA codons are brought together in an orderly linear and sequential fashion. As each new codon is exposed, the appropriate charged tRNA species is bound, and a peptide bond is formed between the amino acid carried by the tRNA and the existing nascent protein chain. The growing chain is transferred to the new tRNA in the process, so that it is held in place as the next tRNA is brought in. Upon completion of translation, the polypeptide chain is released into the cytosol for further processing by other structures, such as endoplasmic reticulum and the Golgi apparatus, followed by association of the newly completed chain with other subunits to form complex multimeric proteins (e.g. hemoglobin), for binding to cofactors, for processing (e.g. glycosylation) in microsomes, etc.

Gene regulation
Virtually all cells of an organism receive a complete copy of the DNA genome transmitted to the organism at the time of its conception. Clearly, the panoply of distinct cell types and tissues found in a complex organism such as *Homo sapiens* is possible only because different portions of the genome are selectively expressed or repressed in each cell. By extension, each cell must 'know' which genes to express, how actively to express them, and when to express them. This biological necessity has come to be known as 'gene regulation' or 'regulated gene expression'. For example, one must understand gene regulation in order to know how pluripotent stem cells determine that they will express globin genes only in those daughter progenitor cells that differentiate along the erythroid lineage, while myeloperoxidase will be expressed in the granulocytic lineage, and platelet glycoproteins in the megakaryocytic lineage.

It has become clear in recent years that major hematologic disorders, such as the leukemias and lymphomas, immunodeficiency states, and myeloproliferative syndromes, result from derangements in the system of gene regulation. An understanding of the ways that genes are selected for expression thus remains one of the major frontiers of biology and medicine. Although little is known about the mechanisms regulating gene expression, many of the molecular elements involved in regulation are becoming better understood as a result of the application of recombinant DNA technology.

Most of the DNA in living cells is inactivated by being bound into a nucleoprotein complex called chromatin. The histone and non-histone proteins in chromatin effectively 'hide' the majority of genes from enzymes needed for expression. It is now clear that there are DNA sequence regions, usually flanking the actual 'structural gene' (the structural gene is defined as the DNA sequence encoding the mRNA precursor), which serve as regulatory signals. These sequences do not usually encode RNA or protein molecules. Rather, they interact with nuclear proteins. These proteins alter conformation of the gene within chromatin in such a way as to facilitate or inhibit access to the apparatus that transcribes genes into mRNA. As noted in Dr Ginder's chapter (Ch. 8), these interactions often 'twist' or 'kink' the DNA in such a way as to reduce the degree to which that particular DNA region is hidden from other molecules by the chromatin proteins. For example, when exogenous nuclease enzymes are added in small amounts to nuclei, these exposed sequence regions are especially sensitive. Thus, 'nuclease hypersensitive' sites in DNA have come to be appreciated as markers for regions in the vicinity of genes that are interacting with regulatory nuclear proteins.

Several types of DNA sequence elements have been defined according to the presumed consequences of their interaction with nuclear proteins. *Promoters* are found just 'upstream' (to the 5' side) of the start of mRNA transcription (the 'CAP') in almost every gene. Promoters appear to be the sequence loci at which mRNA polymerases bind and gain access to the structural gene sequences downstream. They appear to serve a dual function of binding the mRNA polymerase and 'marking' for the polymerase the point at which mRNA transcription should start. *Enhancers* are DNA sequences that serve more complicated and less well understood functions. Enhancers can occur on either side of a gene, or even within the gene in introns. Enhancers appear to bind to nuclear proteins or 'transcription factors' and thereby stimulate expression of genes nearby. The domain of influence of enhancers, i.e. the number of genes to either side whose expression is stimulated by the enhancer, varies. Some enhancers influence only the adjacent gene; others seem to play a central role in marking the boundaries of large multigene clusters (gene domains) whose coordinated expression is appropriate to a particular tissue type or a particular time. Presumably, the nuclear factors interacting with these enhancers are induced to be synthesized or activated as part of the process of differentiation. 'Silencer' sequences appear to serve a function that is the obverse of enhancers. When bound by the appropriate nuclear proteins, silencer sequences cause repression of gene expression.

Assays for detecting nuclear proteins that exhibit 'gene specific' DNA binding are just beginning to achieve widespread utility. There is little information available about these proteins or their biochemical properties. However, proteins involved in the regulation of a few model gene systems have been isolated and their genes cloned. Progress in this area should thus be rapid in the coming years.

BASIC TENETS OF RECOMBINANT DNA TECHNOLOGY

The foregoing discussion makes apparent the fact that the informational content of DNA molecules resides in the nucleotide sequence, rather than in the sugar phosphate backbone. Unfortunately, traditional methods of biochemical fractionation do not provide straightforward means for distinguishing nucleic acid molecules from one another on the basis of their nucleotide sequences. Even if such methods were available, the quantity of bulk genomic DNA necessary to isolate a gene of typical size (a few thousand or tens of thousands nucleotides long) from a complex genome, such as the human genome (three billion base pairs long) renders these methods impractical. In addition, genes do not exist in cells as discrete DNA molecules; rather, thousands of genes are linked together in tandem with very long stretches of intergenic DNA to form chromosomes. For example, in the human genome, the three billion base pairs of DNA exist as 23 chromosomes in the haploid genome. Each chromosome is thus about 100 million base pairs long. These facts render DNA an almost unworkable substance for direct physical purification of most genes.

Recombinant DNA technology circumvents the biochemical problems inherent in the properties of DNA by combining enzymologic, microbiological, and genetic approaches. In this section, I shall merely outline the basic principles from which the strategies are developed so that the reader can better appreciate the more detailed descriptions given in each chapter.

Restriction endonucleases

A major advance in our ability to manipulate DNA molecules was the discovery of enzymes, produced by bacteria, called restriction endonucleases. Restriction endonucleases have the capacity to recognize short nucleotide base sequences (oligonucleotide sequences) and to cleave DNA within or near the recognition sequence. For example, Eco RI, a restriction endonuclease isolated from *Escherichia coli*, cuts DNA at the sequence 5'-GAATTC-3', but nowhere else. Thus, each DNA sample will be reduced reproducibly to an array of smaller sized fragments whose size ranges depend on the distribution with which 5'-GAATTC-3' is encountered. However, the DNA will not be degraded in any other way by the enzyme. Restriction endonucleases differ from other nucleases by the specificity and limited manner with which they degrade DNA.

Part of the 'jargon' that newcomers to the field of recombinant DNA technology often find difficult arises from the shorthand notation given to restriction endonucleases. These enzymes are generally named after the bacterium from which they were isolated. Thus, a restriction endonuclease activity purified from *Serratia marcescens* is called Sma I, that from *Diplococcus pneumoniae* is called Dpn I, etc. Each of the nearly 500 restriction endonucleases that have been described recognizes a unique oligonucleotide sequence and cleaves the DNA only at or near those points. (Table 2.1 shows the names and recognition sites of some typical restriction endonucleases.)

In some cases, different restriction enzymes from different bacterial species have been found to cut DNA at exactly the same recognition sequence. Such restriction enzymes are called isoschizomers. A useful type of isoschizomer is a pair of restriction enzymes which recognize the same sequence but cut or fail to cut according to modifications of the DNA bases, notably methylation. For example, both Hpa II and Msp I recognize the sequence 5′-GCCG-3′. Msp I cuts whether or not the C residues are not methylated, but Hpa II will only cut if the C residues are not methylated. As discussed by Dr Ginder in Chapter 8, these paired enzymes are useful for recognizing those positions in mammalian genomes that are methylated, a modification that has been associated with altered gene expression.

The biological function of restriction enzymes remains poorly understood in microorganisms. It is thought that organisms producing these enzymes use them to degrade unwanted DNA being introduced from outside sources. However, the host DNA frequently contains abundant sites recognized by the endonucleases it produces. Other factors must modulate the biological behavior of the enzymes.

Restriction enzymes have proved to be extraordinarily useful gifts from the microbial world to molecular geneticists. They allow one to reduce the sizes of DNA fragments in a controlled and reproducible manner from several hundred million base pairs long to fragment arrays ranging from a few dozen to a few tens of thousands of bases long. These ranges are far more workable in the test tube. Moreover, by 'mixing and matching' combination of restriction enzymes used to digest the same DNA sample, one can construct maps or 'fingerprints' of the restriction endonuclease sites in a genome. This strategy has made restruction endonuclease digestion as useful an approach for characterizing the fine structure of genomes as proteolytic digestion has been for peptide fingerprinting used by protein chemists. The use of these methodologies is discussed in Chapter 2 and Chapter 8.

As noted by Drs High, Ware, and Stafford in Chapter 2, some restriction endonucleases cut the DNA so as to leave short single stranded overhanging regions or 'sticky ends' at the 5′ or 3′ end of the cutting site, while cleavage by others leaves blunt or flush double stranded ends. Since many restriction endonuclease sites are palindromes (reading exactly the same in forward direction (5′ → 3′) on one strand and the 'backward' (5′ → 3′) direction on the opposite strand, (e.g. Eco RI: 5′-GAATTC-3′ and 3′-CTTAAG-5′)), enzymes leaving overhanging ends are particularly useful. If one digests DNA from two different sources, such as a bacteriophage preparation and a human genomic preparation, with a restriction endonuclease leaving overhanging or 'sticky ends', those ends will be complementary by Watson-Crick base pairing and can thus be annealed together by means of the single stranded overhangs. This is one popular method for generating recombinant DNA molecules.

Enzymes useful for modifying DNA
Several other nucleic acid modifying enzymes have been critical to the development of recombinant DNA technology. Most notable among these are reverse transcriptase (RNA dependent DNA polymerase) and DNA ligase. As noted by Dr Rado and co-authors in Chapter 4 and Dr Mercer in Chapter 7, reverse tran-

scriptase in the enzyme packaged inside retroviruses, which have an RNA genome. In order for retroviruses to reproduce themselves within their cellular hosts, their RNA genomes must be transcribed into DNA molecules (RNA → DNA) that can then be replicated (DNA → DNA) and expressed by host cell machinery (DNA → RNA).

Reverse transcriptase has the very useful property that, if provided with an appropriate 'primer' complementary to a messenger RNA molecule, it can read the mRNA strand in a $3' \to 5'$ direction and transcribe a single stranded DNA copy ('copy DNA', 'complementary DNA', or cDNA) of the RNA molecule. One can thus incubate reverse transcriptase with messenger RNA isolated from a cell or tissue of interest with reverse transcriptase, and generate thereby a family of single stranded DNA molecules representing the entire array of messenger RNAs expressed in that cell or tissue. Using additional enzymes that have been characterized and purified, for example *E. coli* DNA dependent DNA polymerase I, (Klenow fragment), one can synthesize a complementary second strand of DNA (sDNA) from the single stranded cDNA template. This creates a double stranded DNA molecule containing the sequence information originally expressed in the form of mRNA. These DNA molecules can then be manipulated in essentially the same ways that native genomic DNA molecules can, by restriction endonuclease digestion, radioactive labeling, or insertion into microbial host vectors for cloning.

DNA ligase is an enzyme that can join two DNA molecules together to form a single novel DNA molecule. For example, one can join the aforementioned double stranded cDNA molecules with bacteriophage DNA molecules by incubating DNA from both sources together in the presence of DNA ligase (see Ch. 7). This ability to generate artificially recombined or 'recombinant' DNA molecules has given rise to the term recombinant DNA technology.

A number of other important enzymes, discussed throughout the chapters of this volume, have also been useful to the development of recombinant DNA technology. These include a variety of polymerases, kinases, endo- and exonucleases that are used to introduce radioactive residues into DNA molecules, to synthesize new strands, to elongate the ends of DNA molecules by adding single stranded overhanging sequences, to truncate or 'trim' single stranded overhangs in order to generate blunt ended molecules, etc. A vast array of elegant methods have evolved that allow one to modify DNA molecules with exquisite precision. A frequently under-recognized biochemical contribution to the development of molecular genetics has been the ability of many commercial vendors to supply highly purified and active forms of these enzymes as well as highly radioactive and reliable nucleotide triphosphate substrates for incorporation into or addition onto these DNA molecules.

Microbial genetics and infectious DNA molecules

The development of enzyme sources that fragment DNA in a controlled fashion, polymerize it, modify it, or ligate two DNA molecules from dissimilar sources together represents an impressive advance. However, the general utility of these tools would have been limited except for the discovery of certain small and simple DNA molecules that possess remarkable biological properties. Microbial geneticists found that many bacteria harbored DNA molecules that were not part of the single major bacterial chromosome. These novel DNA molecules were found to be rather small (a few thousand to about 100 thousand bases long), to have circular structures, and to carry sequences serving as independent origins of DNA replication. They could thus replicate in host cells independent of the host genome by utilizing cellular DNA replicating enzymes. These DNA molecules can be thought of as elemental commensal organisms, residing in the cell and capable of infecting other host bacteria. They have come to be called extrachromosomal elements or episomes.

The forms of episomes most relevant to this discussion are the plasmid and bacteriophage. Plasmids are circular DNA molecules with the properties noted above. Plasmids useful in recombinant DNA technology usually carry one or more antibiotic resistance genes, an origin of DNA replication, and a limited but useful array of restriction endonuclease sites. Molecular biologists have engineered plasmids with a variety of desirable properties by cutting them with restriction endonucleases and ligating them together in novel combinations. There are a wide variety of useful plasmid vectors available for customized recombinant DNA applications. Typically, these plasmids are only 3000–10 000 bases long. They usually carry genes for ampicillin or tetracycline resistance, as well as a short DNA sequence, containing several tightly clustered restriction endonuclease sites, called a polylinker. The polylinker sequence is inserted into any one of several non-critical regions of the plasmid genome, so that one can utilize the restriction enzyme sites in the linker to open the plasmid for insertion of a genomic DNA or cDNA fragment. Cells 'infected' with these plasmids can be detected and purified by their ability to grow in media containing the relevant antibiotic.

The most useful plasmids for recombinant DNA work are those in which the plasmid or its polylinker includes several restriction endonuclease sites that occur only once in the plasmid genome. Digestion of a circular molecule with an enzyme that makes only a single cut in the circle will cause opening or linearization of the circle while leaving all of the biologically critical sequences intact. One can then insert a DNA molecule into the opening, reseal the circle with DNA ligase and thereby generate a recombinant DNA molecule retaining the biological activities of the parent plasmid.

Bacteriophage are viruses capable of infecting specific strains of bacteria. Their genomes are somewhat more complex than plasmids and the DNA is covered during the extracellular part of the viral life span by a proteinaceous coat. However, bacteriophage genomes relevant to this discussion can also exist in the cell as episomes. The most useful bacteriophage for molecular genetics experiments have been bacteriophage λ, which can be used as a gene cloning vehicle (see Ch. 7), and the single stranded bacteriophage M13, which has proved to be useful for DNA sequencing applications (see Ch. 9). By analogy with plasmid genomes, bacteriophage genomes have been engineered to provide a number of experimentally useful vectors.

The essential aspect of bacteriophage and plasmids important for this discussion is that they are biologically active, even when they exist as simple freestanding DNA molecules. By combining the ability to recombine episomal DNA with DNA from mammalian sources (via restriction enzymes and ligase), with the capacity of the episomes for infection and phenotypic alteration of host cells, one can use these molecules to introduce 'foreign' DNA into host bacteria. Then, all of the useful properties of the vast array of microbial strains available become accessible for the study of genes from other species. Individual strains of bacteria can be readily isolated as single cell clones, grown in extremely large quantities for relatively little expense, and used as 'factories' for the production of the foreign DNA sequence contained within it, as well as its protein products.

Advances in nucleic acid chemistry
During the past two decades, anhydrous methods for the synthesis of DNA molecules *in vitro* have been developed and automated. This has provided a capacity to synthesize short but useful DNA molecules even without the availability of a template or a DNA modifying enzyme. Thus, the polylinker sequences used to introduce restriction endonuclease sites into plasmids can now be readily synthesized by machine and ligated into a plasmid in order to alter its restriction endonuclease map. As noted in Chapters 4 and 11, synthetic oligonucleotides can also be radiolabeled and used as customized molecular hybridization probes.

The tendency of DNA and RNA molecules to prefer existence in double stranded form in physiologic solution has been exploited by nucleic acid chemists for the development of 'molecular hybridization' assays. If DNA or RNA molecules are heated or exposed to certain denaturants, such as formamide, the hydrogen bonds holding two strands together are disrupted and the molecule is denatured into single stranded form. Temperature, salt and denaturant conditions that favor reannealing into the double stranded form can then be restored. This reannealing process is often called molecular hybridization: reannealing under a given set of conditions of temperature, salt, and denaturant is a function of the time of incubation and the concentrations of the two complementary strands.

When DNA and RNA strands are denatured into single stranded form, they will reanneal only with strands having a sequence complementary by the rules of Watson-Crick base pairing. Thus, one can denature a specimen of DNA or RNA (for example, messenger RNA from a human reticulocyte) and incubate it with a radioactively labeled, defined DNA or RNA sequence, for example, a cloned human β-globin gene. During the reannealing reaction, the labeled β-globin gene DNA probe will hybridize only to those mRNA molecules that are complementary by Watson-Crick base pairing, i.e. β-globin messenger RNA molecules. One can then utilize any one of several available techniques to recognize or separate the fraction of radioactively labeled DNA 'probe' molecules that have been bound into double stranded form from the unbound or unreacted single stranded molecules. For example, the enzyme S_1 nuclease degrades single stranded DNA molecules, leaving only the double stranded 'hybridized' molecules intact. The result is a convenient assay for detecting and quantitating the globin messenger RNA within the reticulocyte mRNA preparation. By extension of this reasoning, one can use molecular hybridization strategies to detect, quantitate, and map DNA sequences or RNA sequences derived from any tissue or source for which a complementary defined DNA probe is available.

Numerous variations of the basic molecular hybridization have been devised for a wide variety of applications, as will be apparent from the chapters in this volume. Detailed discussion is beyond the scope of this chapter. However, the range of applications, theoretical rationale, and utility of most of these assays can be appreciated by their analogy to the use of antigen/antibody reactions in immunochemistry. The DNA probe serves the molecular geneticist in much the same way as a defined antibody probe serves the immunologist. The principles underlying the various molecular hybridization techniques used are very similar to those devised for using antibody probes to quantitate and detect defined antigens.

Coalescence of methodologies to produce and isolate recombinant DNA molecules

Advances in each of the areas just noted have been brought together for the purpose of physically isolating genes from complex genomes, such as a mammalian genome. Detailed methods for gene cloning are discussed in Chapters 5, 6, and 7. In this introduction it is sufficient to outline the basic algorithm. DNA is isolated from, for example, a mammalian cell preparation, and digested with restriction endonucleases to generate overhanging 'sticky ends'. (Alternatively, if one wishes to isolate only those DNA sequences encoding the specific array of genes expressed in a given tissue, one first isolates messenger RNA and converts it into cDNA by incubation with reverse transcriptase as a first step.) An infectious plasmid or bacteriophage DNA molecule is cut with the same restriction enzyme so that molecules from the two sources have complementary 'sticky' ends. The DNAs from the two sources are incubated together in the presence of DNA ligase under conditions of a slight excess of the microbial DNA molecules. This ensures that each plasmid or bacteriophage DNA molecule ligates to only one molecule from the mammalian source. The recombinant DNA molecules are then 'sealed' with DNA ligase. Each recombinant molecule is thus an infectious DNA species carrying a single DNA fragment from the mammalian source as a 'passenger'.

The recombinant DNA molecules are then used to infect an excess of host bacterial cells. The use of an excess number of cells ensures that each cell, on the average, acquires only a single recombinant DNA molecule. The host cells are chosen to lack some phenotypic property conferred by the infecting molecule, such as antibiotic resistance. The infected cells are then 'plated' on petri plates at a density allowing detection of individual colonies or bacteriophage plaques. Each colony or plaque represents the progeny of a single cell, and is thus a 'clone' of cells or phage carrying a single DNA fragment from the mammalian source. Therefore, that DNA fragment, or gene, has been physically and genetically isolated in its host cell from all other mammalian DNA fragments by the cloning process.

What remains is the need to identify the DNA fragment representing the specific gene one wishes to purify. One must identify within the array of plaques or colonies, called a recombinant DNA 'library', those cells or phages carrying the DNA sequence of interest. Numerous devices have been developed for screening these libraries for the presence of the occasional clone bearing the gene of interest. As discussed in Chapter 6 and Chapter 7, different approaches are suitable depending on what information is available about the particular gene or its protein product. Discussion of these detailed areas is beyond the scope of this chapter.

Once one has identified the colony or bacteriophage plaque containing the recombinant molecule of interest, that colony or plaque can be separated from the remainder of the library and amplified by growing in bacterial culture. In this manner, one can produce substantial quantities of recombinant DNA molecules from the cloned host cell. With respect to other DNA molecules derived from the same mammalian source, the recombinant DNA 'cloned' gene will be absolutely pure. The purified gene can then be used as a hybridization probe, as the substrate for obtaining its DNA sequence, or as a template for controlled expression and production of its mRNA and protein products.

The elegance of recombinant DNA technology resides in the capacity it confers upon investigators to examine each gene as a discrete physical entity that can be purified, reduced to its basic building blocks for decoding of its primary structure, analyzed for its patterns of expression, and perturbed by alterations in sequence or molecular environment so that the effects of changes in each fine structural region of the gene can be assessed. Moreover, techniques have been developed, as discussed in Chapter 11, whereby the purified genes can be deliberately modified or mutated to create novel genes that are not, in many cases, available in nature. These provide the potential to generate useful new biological entities, such as modified viruses that can serve as vaccines, modified proteins customized for specific therapeutic or industrial purposes, or altered combinations of regulatory and structural genes that allow for the assumption of new functions by specific gene systems. For example, it is now theoretically possible to utilize bone marrow cells (which can be readily removed from an individual, and later transplanted back into that individual or a compatible patient) as vehicles for gene therapy. It is theoretically possible to insert a gene for an important serum protein, such as a clotting factor, into the explanted cells then to reimplant the bone marrow cells into the patient. The genetically engineered bone marrow cells will then serve as surrogate sources of a protein for which a patient may be deficient, even if the 'natural' cellular production of that protein is some other less accessible tissue, such as endothelial cells.

The specific ways by which purified genes greatly strengthen the arsenal used by molecular biologists to attack the mysteries of gene regulation can be summarized as follows:

First, the abundant and pure DNA fragments made available by molecular purification of a gene provide the characterized DNA sequences needed for use as hybridization probes in molecular hybridization assays. Thus, a purified gene is the source of molecular hybridization probes needed for the Northern or Southern blotting assays described in Chapters 2 and 3. Cloned globin genes can then be used for analysis of globin genes and mRNAs in specimens from normal patients or patients with hemoglobin disorders.

Second, cloned genes allow one to accumulate sufficient amounts of a pure homogenous DNA sequence for determination of the exact nucleotide sequence of the gene, as described in Chapter 9. Indeed, DNA sequencing techniques have become so reliable and efficient that it is frequently far easier to clone a gene encoding a protein of interest and determine its DNA sequence than it is to purify the protein and determine its amino acid sequence. As noted earlier, the DNA sequence of the gene will predict exactly what the amino acid sequence of the protein must be, since that amino acid sequence is encoded in the gene DNA. By comparing normal gene sequences with the sequences of genes cloned from patients known to have abnormalities of a specific gene system, such as the globin genes in the thalassemia or sickle cell syndromes, one can compare the normal and pathologic 'anatomy' of genes critical to major hematologic processes. In this manner it has been possible to identify over 100 mutations responsible for various forms of thalassemia, hemophilia, red cell enzymopathies, porphyrias, etc.

Third, each purified cloned gene can be further manipulated by extension of the same types of 'cutting and pasting' techniques just described for studies of gene expression. Just as plasmid and bacteriophage vectors have been developed for the transfer of genes into microbial host cells, a variety of means for reasonably efficient transfer of genes into eukaryotic cells have been determined. One set of relevant strategies is described in Chapter 13. By judicious and adept application of these gene transfer technologies, one can place the gene into a controlled cellular environment and analyze the expression of that gene. These 'surrogate' or 'reverse' genetics systems permit one to analyze the normal physiology of expression of a particular gene, as well as the pathophysiology of abnormal gene expression resulting from mutations.

Fourth, detailed knowledge about the structure and expression of cloned genes greatly strengthens the opportunities investigators have to examine their protein products. By expressing the cloned genes in large amounts in microorganisms or eukaryotic cells, one can produce customized regions of proteins for use as immunogens, thereby allowing preparation of a variety of useful and powerful antibody probes for direct study of the protein products. Alternatively, one can prepare synthetic peptides deduced from the DNA sequence for use as immunogens. Controlled production of large amounts of the protein also allows one to conduct direct analysis of specific functions attributable to regions within that protein.

Finally, all of the above techniques can be extended greatly by taking advantage of methods available for mutating the genes and examining the effects of those mutations on the regulated expression of the genes and the properties of mRNA and proteins encoded by them. By 'swapping in' portions of one gene within another (chimeric genes), or abutting structural regions of one gene with regulatory sequences of another, one can investigate in previously inconceivable ways the complexities of gene regulation. These 'activist' approaches to modifying gene expression create the opportunity to generate new RNA and protein products of genes whose applications are limited only by the collective imagination of molecular biologists.

The most important impact of the genetic approach to the analysis of biological phenomena is presently the most indirect. Diligent and repeated application of the above algorithm to the study of many genes from diverse groups of organisms is beginning to reveal the basic strategies used by nature for the regulation of cell and tissue behavior. As our knowledge of these 'rules of regulation' grows, our ability to understand, detect, and correct pathologic phenomena will increase massively. Similarly, our capacity to use biological strategies for production of useful substances is currently limited largely by our limited knowledge of the factors that constrain genes from maximal expression in various cell types. As our knowledge of these natural rules improves so too will our opportunities to engineer useful biologicals.

BASIC LABORATORY HYGIENE FOR STUDY OF NUCLEIC ACIDS

It has been this author's experience that many highly capable investigators view the possibility of adopting recombinant DNA technology with a rather puzzling sense of apprehension. Recombinant DNA technology appears to be a formidable skill to acquire because it demands the simultaneous use of so many different types of methodologies: enzymologic, microbiological, tissue culture, and nucleic acid chemistry. It is this author's hope that the reader will be reassured by the relative simplicity of the actual methods described in this book. The techniques can be and have been so widely used for study of virtually every system in biology that they have been very well standardized. Highly reliable reagents have become available from commercial sources, often in the form of kits. Moreover, as a quick survey of the following chapters will reveal, each individual method is relatively simple.

Compared to many other techniques widely utilized in biological research, the individual methods comprising recombinant DNA technology require relatively little in the way of specialized equipment, unusual dexterity, or special theoretical skills. The key to success with recombinant DNA technology is the ability to organize and maintain several sets of methods in good working order simultaneously so that they can be reliably recruited to the experiment at the appropriate time, 'on demand'. In that regard, a major challenge lies in the fact that many of the starting materials used as 'off the shelf' reagents are themselves very fragile and complex macromolecules such as enzymes, plasmid DNA preparations, etc. Most failures at the methodologic level in molecular biology laboratories can be traced to the fact that inadequate attention is often paid to the highly fragile nature of these reagents and the 'TLC' with which they must be handled. Therefore, the following 'rules' of laboratory hygiene serve as an underpinning to each of the chapters that follow.

First, most enzyme preparations and many nucleic acid specimens are either thermolabile, or exquisitely susceptible to degradation by contaminating proteases or nucleases at wet ice temperatures (0–5°C), room temperature, or water bath incubator temperatures typically used in biological laboratories (30–60°C). Enzymes should be stored at −20 or −70°C as recommended by the manufacturer, and should never be subjected to freeze thaw cycles, since these inactivate many preparations. If enzymes need to be diluted, the diluent should include sufficient glycerol to permit storage at the above temperatures without freezing damage. Radioactive nucleotides are similarly best preserved at low temperatures. If specimens must be thawed, they should be aliquoted into appropriately small sized samples at the time of the first thawing. Individual aliquots should then be refrozen and thawed when needed. In general, that is the maximum number of freeze cycles that should be tolerated.

Enzymes should be exposed to temperatures above their optimal storage temperatures for only the minimum necessary periods of time. Many widely advertised 'freezing blocks' are now available for holding enzymes at −20°C during the assembly of reaction mixtures. These allow one to bring enzymes to the bench without raising their temperature. Alternatively, it is often a good practice to assemble the rest of the reaction mix and bring it near to the freezer so that the enzyme is removed from storage only long enough for the actual pipetting.

Loss of enzymes and nucleic acid preparations by contamination of samples with degradative enzymes is one of the most common disasters that befall molecular biology experiments. Several rules should be religiously followed in order to avoid this complication. It should be remembered that in any laboratory utilizing biological tissues, lysosomal nucleases and proteases released during manipulations of those tissues are essentially ubiquitous in the environment. The environment must thus be excluded from vessels containing precious macromolecules.

One should always wear gloves when handling these substances, and change gloves as frequently as feasible so that material is not carried from potentially nuclease or protease rich tissues to purified macromolecular preparations. As noted by Dr Leibowitz in Chapter 3, either sterile glassware or clean plastic ware should be employed exclusively. All reagents should be either autoclaved or treated to remove offending enzymes by agents such as diethylprocarbonate. Stock tubes containing enzymes or plasmid preparations should never be entered with any pipet or transfer device that has touched any other specimen. One should use disposable pipet tips, and enter a stock preparation only with a previously unused tip. This is important not only to avoid nuclease contamination, but also to avoid cross contamination with potentially confounding agents. For example, if one contaminates a stock tube of Eco RI restriction enzymes with even a small amount of another restriction enzyme, such as Hind III, one can obtain impossibly confusing restriction endonuclease results. Inadvertent contamination of cellular DNA specimens with cloned genes residing on plasmids has misled many an investigator into thinking he or she was pursuing a biological phenomenon that was really a simple artifact. These catastrophes can be avoided simply by adherence to the common sense rules of 'cleanliness' and organization outlined above.

Radioactive isotopes, notably ^{32}P and ^{35}S, remain essential for tagging DNA, RNA, and protein molecules in molecular biology laboratories, even though non-isotopic labeling methods are being developed at a gratifying pace. Each of these isotopes is potentially dangerous. Any laboratory intending to pursue molecular biology experiments should take pains to establish all precautions for the handling of radioisotopes, such as lead and plexiglass shielding, and maintain a close working relationship with local radiation safety officials.

Most research funding agencies and institutions require investigators utilizing recombinant DNA molecules to register their experiments with appropriate advisory and regulatory bodies. Some experiments employing potentially dangerous DNA molecules, such as cloned hepatitis viruses, AIDS virus, oncogenic viruses, etc. require certification of particular experiments to ensure that they are conducted in an appropriate biohazard containment facility. Thus, before embarking on the methods described in this book, each investigator is urged to contact the Recombinant DNA Advisory Committee constituted at his or her institution.

2
Isolation of genomic DNA, restriction endonuclease mapping and Southern gene blotting

K. A. High   J. L. Ware   D. W. Stafford

BACKGROUND

The elucidation, based on X-ray diffraction data, of the structure of DNA by Watson and Crick in 1953, and the subsequent 'breaking' of the genetic code constitute the fundamental theoretical basis of modern molecular biology. To a great extent, however, the recombinant DNA techniques which have been responsible for the explosion of knowledge in the biological sciences rested on technical advances in other fields that occurred at about the same time. Progress in nucleic acid chemistry permitted straightforward and reliable isolation of DNA and RNA; advances in microbial genetics provided a variety of restriction enzymes which made it possible to manipulate and analyze DNA; and work in virology and microbiology provided access to several other key nucleic acid modifying enzymes such as reverse transcriptase and DNA polymerase. Two of these techniques — DNA isolation and restriction enzyme mapping — form the basis of this chapter. Southern blotting, which utilizes both of these techniques to allow one to determine the structure of a particular gene, is also discussed here.

ISOLATION OF HIGH MOLECULAR WEIGHT DNA

High molecular weight eukaryotic DNA is the starting point for construction of genomic libraries in phage or cosmid vectors. In addition, DNA isolation is required for the analysis of the presence and number of specific sequences by Southern blotting. The general outline of all DNA isolation procedures is the same (Fig. 2.1). First, cells are disrupted by homogenization in a detergent, and nuclei are pelleted by centrifugation. Next, the nuclei are resuspended and treated with an ionic detergent such as sodium dodecyl sulfate (SDS), which dissociates the deoxyribonucleoprotein complex and denatures nucleases. Protein is then removed by digestion with a powerful proteolytic enzyme, proteinase K, which is active in the presence of SDS. The degraded protein and proteinase K are subsequently removed by phenol/chloroform extraction. The nucleic acids, which remain in the aqueous phase, are treated with RNase, and the DNA is recovered by ethanol precipitation. DNA can then be quantitated either by UV spectroscopy or by estimation from gel electrophoresis with standards of known concentration.

22 MOLECULAR GENETICS

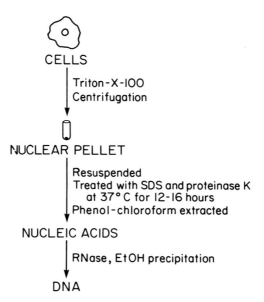

Fig. 2.1 Schematic diagram of procedure for isolation of high molecular weight DNA. See text for details.

Modifications of the procedure are required depending upon the cells of origin (blood or cells in suspension culture vs. solid tissue) or the size of the DNA required (e.g. very high molecular weight DNA, > 50 kb should be isolated by dialysis rather than ethanol precipitation). The following detailed procedure is one which we have used extensively and which provides good yields of high MW DNA (250–500 μg from 10 ml blood).

This procedure is adapted from that of Blin & Stafford and is for blood (i.e. DNA from WBC nuclei). One volume of anticoagulated (citrated) blood is mixed with 9 volumes of 0.32 M sucrose/10 mM Tris HCl (pH 7.5)/5 mM $MgCl_2$/1% Triton X-100 at 4°C. Nuclei are pelleted by centrifugation at 1000 g for 10 minutes and then suspended in 0.45 × original blood volume of 0.075 M NaCl/0.024 M EDTA, pH 8.0. SDS is added to 0.5%, and proteinase K to 2 mg/ml, and the mixture incubated at 37°C overnight. The digest is then gently mixed with an equal volume of phenol (saturated with 20 mM Tris HCl, pH 8.0), followed by addition of an equivalent volume of chloroform/isoamyl alcohol (24:1). The phases are separated by centrifugation at 1000 g for 15 minutes. The aqueous (upper) phase is removed and re-extracted with chloroform/isoamyl alcohol. After centrifugation, the aqueous phase is removed and pancreatic RNase (preheated by incubating at 80°C for 20 minutes to destroy DNase activity) is added to 50 μg/ml, and incubated at 37°C for 2 hours. The mixture is re-extracted with phenol:chloroform, then chloroform:isoamyl alcohol, and DNA is precipitated by adding 0.1 volume of 3 M sodium acetate (pH 5.2) and 2.5 volumes of cold 100% ethanol. The DNA will precipitate immediately and can be removed by swirling gently with a Pasteur pipet (spooling) and redissolved in TE (10 mM Tris HCl, pH 7.5/1 mM EDTA) or can be stored in ethanol indefinitely at −20°C. DNA stored in ethanol can be recovered by centrifugation at 10 000 rpm for 10–15 minutes. The pellet is then washed with 100% ethanol to remove salt. DNA, especially smaller fragments, is soluble in ethanol in the absence of salt, so that it is best not to wash extensively. When dealing with small pellets, we remove the aqueous ethanol layer with a drawn-out Pasteur pipet, taking care not to dislodge the pellet.

When obtaining blood samples for DNA preparation, it is best to avoid the use of heparin as an anticoagulant, since even small amounts of heparin can inhibit the activity of restriction enzymes. We have included two procedures (meta-cresol precipitation and 2-methoxyethanol extraction, *vide infra*) for removing heparin from blood samples, but both of these result in substantial decrease in DNA yield. It is considerably better to avoid the problem altogether by anticoagulating samples with EDTA or citrate.

Citrate or EDTA anticoagulated blood samples which have been stored at 4°C (for up to several days) can be used to prepare DNA. The major potential difficulty with stored samples is the possibility of degradation of DNA by endogenous nucleases. Storage at low temperatures (0–4°) decreases the activity of these nucleases, and storage at −70°C virtually abolishes it. Thus, if long-term storage of tissues is required prior to DNA extraction, it is best done at −70°C.

Minor modifications of the above procedure allow the efficient isolation of high molecular weight DNA from solid tissues.[1] The tissue is minced with scissors and quick-frozen in liquid nitrogen. The frozen tissue is pulverized in a Waring blender; the liquid nitrogen is allowed to evaporate and the resulting powder is suspended in 9 volumes of lysis buffer as described in the foregoing procedure. From this point the protocol continues as described.

For most analytical procedures, high molecular weight DNA (> 50 kb) is preferred. For some techniques, for example, cloning into cosmids, it is essential. To obtain high molecular weight DNA ethanol precipitation should be avoided. Instead, following phenol:chloroform extraction of RNase treated DNA, the DNA should be dialyzed extensively against 10 mM Tris, 1 mM EDTA at 4°C until the optical density of the fluid outside the bag is less than 0.05 at 270 nm.[2] The size of this DNA can be assessed by running on a 0.4% agarose gel; the average MW should be greater than that of intact λ DNA. It is important to realize that apparent molecular weight in this circumstance can be misleading because of the tendency of very high molecular weight DNA to 'snake' through the gel.

A potential difficulty of this procedure for preparing very high molecular weight DNA is that one generally recovers the DNA in an inconveniently large volume. To reduce the volume, the solution can be extracted several times with 2-butanol. Water molecules will be partitioned into the organic phase, but DNA and other solutes will not. A straightforward protocol for this procedure has been published by one of us (DWS).[3] The procedure is also described in appendix A of 'Molecular cloning: a laboratory manual'.[4]

Isolation of high MW DNA is a straightforward and reliable procedure. The major pitfalls in the procedure are of three types: (1) contamination with e.g. phenol, carbohydrates or proteins, which may inhibit subsequent restriction enzyme (RE) digestion or interfere with accurate quantitation, (2) nuclease degradation, and (3) shear degradation. Phenol and some carbohydrates inhibit the activity of REs. Phenol contamination is readily detected on UV scan by a peak of absorbance at 270 nm, and is removed by repeated chloroform:isoamyl alcohol extractions. Protein contaminants, which can also inhibit the activity of restriction enzymes, can be removed by additional phenol:chloroform extractions. If carbohydrate or polysaccharide (e.g. heparin) contamination is suspected, then the sample can be reprecipitated with meta-cresol[5,6] as follows: to a DNA sample at a concentration of about 0.1–5.0 μg/λ, add an equal volume of 40% sodium benzoate. Next add 0.2 volume of meta-cresol. Vortex and spin down at 10 000 rpm × 10 minutes. Remove the supernatant and wash the pellet in 70% ethanol. Redissolve the pellet in glass distilled water, add 0.1 volume of 3 M sodium acetate and 2.5 volume absolute ethanol and precipitate overnight at −20°C. The pellet can now be spun down out of ethanol, washed once more, and dissolved in the appropriate restriction enzyme buffer. We have used this technique successfully in reprecipitating samples drawn in heparin which initially failed to cut with restriction enzymes. Another procedure which has been reported to be successful at removing heparin, but with which we do not have personal experience, is 2-methoxyethanol extraction.[7] In this procedure, an equal volume of 2.5 M KPO$_4$, pH 8.0, is added to the DNA solution and mixed gently. The same volume of 2-methoxyethanol is then added, followed by gentle mixing and recovery of aqueous phase after centrifugation. Both of these procedures result in some loss of material and are generally not performed unless the sample proves unrestrictable and other contaminants have been excluded.

Deoxyribonucleases require divalent metal ions for their activity; inclusion of EDTA, which chelates these ions, in the lysis buffer is generally sufficient to inactivate them. In addition, they are rapidly digested by proteinase K. Shear degradation is prevented by avoiding vigorous mixing and pipetting, and in addition, if very high MW DNA is required, by avoiding ethanol precipitation.

There are two reliable and widely used methods for quantitating DNA: UV spectrophotometric measurement and ethidium bromide fluorescence. If the sample is abundant and relatively pure, UV spectrophotometry is the method of choice. Readings are taken at 260 and 280 nm. An OD reading of 1 at 260 nm corresponds to a DNA concentration of 50 μg/ml. (Editor's note: RNA is quantitated in a similar fashion. An OD reading of 1 corresponds to about 40 μg/ml of RNA.) A relatively pure preparation of DNA or RNA should display a peak absorbance at OD_{260} and give an OD_{260}/OD_{280} ratio of 1.8. If the ratio is considerably less, contamination with protein or phenol should be suspected. Further analysis at 230 nm should confirm a peak absorption at OD_{260} such that the OD_{260}/OD_{230} ratio is approximately 2.0. An OD_{230} reading higher than the OD_{260} value indicates contamination and prevents an accurate determination at OD_{260}. In this case, or in the case where the quantity of DNA is very small, measurements from the fluorescence generated by the intercalator, ethidium bromide, are more accurate. By comparing the fluorescence in agarose gels of known concentrations of DNA to unknown concentrations, a reasonable quantitation may be obtained. The details of this procedure are outlined in the Maniatis manual (pp 468–469).[4] A barely visible band corresponds to about 100–200 ng of DNA.

RESTRICTION ENZYMES; RESTRICTION MAPPING

Restriction endonucleases are bacterial enzymes which recognize and cut specific sequences in double stranded DNA.[8] In bacteria, they probably serve to protect against phage infection. For the molecular geneticist, they have proved an extremely powerful tool for analyzing and manipulating fragments of DNA. Over 400 restriction enzymes have now been purified and characterized. Type II restriction enzymes recognize a particular oligonucleotide sequence (see Table 2.1) and introduce a cut at the recognition site (in contrast to Type I enzymes, which recognize a site and introduce a cut elsewhere). The cuts made by restriction endonucleases can generate either cohesive (so-called 'sticky') ends, or blunt ended fragments (see Fig. 2.2).

The utility of restriction endonucleases stems from two basic characteristics of these enzymes:

1. Recognition and cutting at a specific sequence means that a particular enzyme produces a unique and reproducible set of fragments following digestion of a particular DNA sample.

2. The generation of compatible termini makes possible the 'cutting and splicing' of heterologous fragments of DNA. Compatible ends can be connected by the use of a DNA-joining enzyme, DNA ligase. Restriction enzymes then

Table 2.1 Commonly used restriction enzymes and their recognition sites.

Name of enzyme	Microorganism	Recognition sequence and cleavage site
Eco RI	*Escherichia coli*	↓ GAATTC CTTAAG ↑
Bam HI	*Bacillus amyloliquifaciens H*	↓ GGATCC CCTAGG ↑
Hind III	*Hemophilus influenzae*	↓ AAGCTT TTCGAA ↑
Alu I	*Arthrobacter luteus*	↓ AGCT TCGA ↑

TYPES OF CUTS MADE BY RESTRICTION ENZYMES

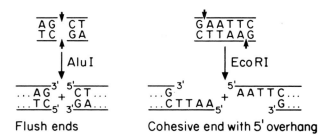

Fig. 2.2 Type II restriction endonucleases. These enzymes recognize and cut specific sequences in DNA. The fragments generated may have blunt ends (Alu I), or 5' (Eco RI) or 3' (not shown) overhangs.

allow the efficient generation of recombinant DNA molecules, e.g. insertion of eukaryotic DNA into a phage or plasmid vector, joining of a metallothionein promoter to a growth hormone çDNA, etc. Thus, the two major purposes of restriction digests include:

1. restriction mapping (*vide infra*) of a particular DNA fragment, and
2. preparation of a fragment for other uses, e.g. cloning into phage, radio-labeling, etc.

The remainder of this section will focus on practical aspects of using restriction enzymes, particularly aspects of the reaction conditions which affect enzyme activity and sequence fidelity.

The components of a restriction digest, in addition to the restriction enzyme itself, include the substrate DNA, a buffering system, a source of Mg^{2+} ions, and NaCl or KCl in varying concentrations. Restriction enzymes are measured in *units*; one widely used definition equates one unit to the amount of enzyme required to digest completely 1 μg of DNA in 1 hour. The substrate DNA in most 'unit' definitions is λ DNA at a concentration of roughly 50 μg/ml. For much more concentrated samples of substrate DNA, the 'unit' definition may not be accurate (*vide infra*). (Manufacturer's definitions of units may vary; one should always check the manufacturer's information before using REs.) In addition to the quality and amount of the enzyme, the two most important factors for determining progress of the reaction to completion are the purity of the substrate DNA, and the time and temperature conditions of the reaction. As discussed earlier, contamination of the DNA substrate by any of a variety of compounds can significantly impede the reaction. For example, proteins may interfere if they bind to substrate DNA. Organic solvents used in DNA isolation (phenol, chloroform) can destroy the restriction enzymes themselves, as will SDS. Contaminating EDTA can chelate the Mg^{2+} required as a cofactor in restriction digests. High concentrations of NaCl (used in a number of procedures for purifying a particular DNA fragment from a gel) can inhibit the reaction or promote 'star' activity (cleavage at sites other than the usual recognition sequence). Thus, great care must be taken to remove these components from the DNA to be cut. On the other hand, contaminating RNA does not interfere with these reactions, nor does contamination with other DNA (although DNA contamination may considerably complicate interpretation of restriction digest results). Finally, if the substrate DNA is very concentrated, the viscosity of the solution can inhibit enzyme diffusion. In practice, this is not usually a problem for DNA substrate concentrations of < 500 µg/ml.

The time required for the reaction to go to completion is inversely proportional to the amount of enzyme used and can be calculated as follows:

$$\mu g \text{ DNA to be cleaved} = \text{units of enzymes} \times \text{hours of incubation}$$

(10 µg of DNA would require 10 units of enzymes for 1 hour, or 5 units for 2 hours). It is common practice to use a 2–3 fold excess of enzyme as calculated from this formula. Although it is theoretically possible to save on enzyme by prolonging incubation times, this practice is not recommended for two reasons: first, many enzymes are not active beyond 1–2 hours after beginning the reaction; and second, if the DNA or the enzyme has even trace amounts of contaminating non-specific nucleases, these will generally survive better than the specific restriction endonuclease and thus result in loss of specificity in the reaction.

Most commonly used restriction enzymes exhibit greatest activity at 37°C, although reactions will proceed, albeit more slowly, at lower or higher temperatures. A few enzymes have temperature optima at 60–65°C. In these cases, the restriction enzyme reaction can be covered with a drop of paraffin oil to prevent evaporation.

In addition to DNA purity, and time and temperature conditions, one must also give some consideration to the other components of the reaction. All restriction enzyme digests must be buffered; since most have pH requirements in the range of 7–8, tris (hydroxymethyl) aminomethane (Tris) is generally a useful buffer system. It is important to adhere closely to the manufacturer's recommendations for pH, since many enzymes exhibit altered activity ('star' activity) if the pH is changed. For example, Eco RI at pH 7.5 recognizes and cleaves the sequence GAATTC; at pH 8.5, however, it loses specificity and may cleave at AATT as well. Other factors which can promote 'star' activity include high enzyme-to-DNA ratios, the presence of organic solvents, and high NaCl concentrations.

Most enzymes are stored in 50% glycerol, which prevents their freezing while stored at −20°C. However, the activity of some enzymes is inhibited by concentrations of more than 5% glycerol in the reaction mixture. Higher concentrations can also promote 'star' activity. Thus, the restriction enzyme itself should never be more than one-tenth of the reaction volume.

Two other components of a usual restriction digest are BSA and dithiothreitol. Bovine serum albumin (Miles Scientific, Pentax fraction) is generally added in concentrations of 10–100 µg/ml. It is said to protect the RE from proteases and from non-specific absorption. Some reactions call for either dithiothreitol or 2-

mercaptoethanol. These should not be added unless they are specifically required since theoretically they may function to stabilize contaminating non-specific endonucleases.

Although it is possible to prepare reactions from stock solutions each time, it is far simpler to prepare a series of 10 × buffers, which may be frozen and then thawed for each use. Again, although one could prepare 10 × buffers for each enzyme, in many laboratories, a small number of core buffers, each appropriate for several enzymes, is maintained. The formulas for these core buffers, along with their corresponding enzymes, are given in Table 2.2. It should be noted that there are several enzymes which cannot be accommodated in this system and for which separate buffers must be prepared. Many commercial enzyme suppliers provide the appropriate 10 × buffer for each purchased enzyme.

Table 2.2 Core buffers for restriction enzymes.

	10 × no salt	10 × low salt	10 × med salt	10 × high salt
NaCl	0	60 mM	500 mM	1000 mM
Tris HCl (7.4)	100 mM	100 mM	100 mM	100 mM
MgCl$_2$	100 mM	100 mM	100 mM	100 mM
BSA	1000 µg/ml	1000 µg/ml	1000 µg/ml	1000 µg/ml
	Hha I[1]	Acc I[1]	Eco RI	Bam HI[2]
	Nar I[1]	Kpn I[1]	Alu I	Bgl II[1]
	Sac I[1]	Xmn I[1]	Ava II[1]	Bst NI[2]
	Sac II[1]		Cla I	Dde I[1,2]
	Xho II[1]		Dra III	Hine II[1]
			Hae III[1]	Mst II[1,2]
			Hind III	Nco I[2]
			Pvu II[1]	Pst I
			Rsa I[1]	San 3A
			Xbu I	Tag I[1]

	10 × low KCl	10 × med KCl
KCl	60 mM	500 mM
Tris HCl (pH 7.5)	100 mM	100 mM
MgCl$_2$	100 mM	100 mM
BSA	1000 µg/ml	1000 µg/ml
	Hpa II[3]	Aat II[3]
	Mbo II[3]	

1 Requires 2-mercaptoethanol or dithiothreitol
2 Final optimal NaCl concentration 150 mM
3 Requires DTT or 2-mercaptoethanol

A typical protocol for setting up restriction digests follows:
1. For a digest of 1 µg of DNA, let the total reaction volume be 20 µl. To a labelled eppendorf tube, add the substrate DNA, 2 µl of 10 × buffer, and enough sterile distilled water to make up the volume to 19 µl. Mix by vortexing.
2. Now add 1–2 units of enzyme (1 µl volume in this example) and mix by gently tapping the tube. The restriction enzyme should not be more than 10% of the total reaction volume. Mix all components of the reaction before adding the restriction enzyme. When the tube is ready, take the enzyme from the

freezer, add it to the reaction, and immediately return it to the freezer. Always use a fresh pipet tip in the enzyme stock.
3. Incubate the tube for 1–2 hours at 37°C.
4. To stop the reaction, add EDTA to 10 mM. Alternatively, if the samples are to be loaded onto a gel, e.g. for Southern blotting, one can stop the reaction by adding a 0.1 volume of loading buffer (20% sucrose, 10% Ficoll, 100 mM EDTA, 1% bromophenol blue).

To carry out restriction mapping one generally makes use of either double digests or partial digests (or both) to define contiguous fragments. The principles of restriction mapping using double digests are outlined in Fig. 2.3. As a practical consideration, note that if the two enzymes have similar reaction conditions, then the digests can be carried out simultaneously. If not, the usual approach is to begin with the enzyme having the lower salt concentration, then after a suitable interval (1–2 hours), to add extra salt and the second enzyme. In some instances it is necessary to add distilled water to the reaction along with the second enzyme, in order to keep the total amount of restriction enzyme less than 10% of the reaction volume. An alternative approach is to extract the first reaction mixture with an equal volume of phenol–chloroform, then precipitate the DNA from the aqueous phase by adding 0.1 volume of 3 M Na acetate (pH 5.2) and 2.5 volumes of absolute ethanol. After 30 minutes at −70°C, the DNA is centrifuged for 10 minutes at 10 000 rpm. The pellet is dried and then made up in the appropriate buffer for the second enzyme.

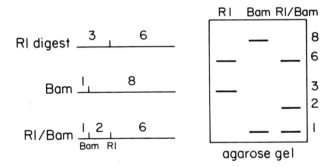

Fig. 2.3 Use of double digests to map DNA fragments. In this example, two enzymes are used to generate a map of a 9 kb fragment. The patterns generated by digestion with a single enzyme are established first. In this case, Eco RI digest of the 9 kb fragment yields a 6 kb and a 3 kb fragment, and Bam digestion yields an 8 kb and a 1 kb fragment. The relative ordering of the restriction sites is made apparent by the double digest, which yields fragments, of 6, 2, and 1 kb. Thus, the Bam site must lie within the 3 kb Eco RI fragment.

Partial digests are also useful in ordering contiguous fragments, since they cut at only a fraction of the possible sites. In a partial digest, the sizes of fragments generated represent sums of the fragments generated by a complete digest. To carry out partial digests, it is often best to use a range of enzyme concentrations on several aliquots of the DNA to be analyzed. The results are monitored by gel electrophoresis.

SOUTHERN BLOTTING

Southern blotting is a technique which allows one to detect and characterize specific DNA sequences. Originally described by EM Southern in 1975,[9] the procedure involves digesting the DNA to be analyzed with a restriction enzyme, separating the resulting fragments by agarose gel electrophoresis, denaturing the DNA *in* the gel and then transferring it to a nitrocellulose or nylon membrane. The membrane is incubated with a radiolabeled probe which is complementary to the DNA to be detected. Following hybridization, the unbound probe is washed off and the position and intensity of the hybridizing band are detected by autoradiography. The technique is used extensively in two settings: first, in characterizing recombinant phage and plasmids, and second, in detecting and mapping sequences in genomic DNA. This latter application requires an extraordinary degree of sensitivity, since one is generally attempting to detect a single copy gene from among an organism's total DNA.

The first step in Southern blotting is restriction enzyme digestion of the DNA to be analyzed. It is important in this instance that complete rather than partial digests be obtained, since the band sizes resulting from a partial digest could confound interpretation of results. In the next step, fragments are separated by gel electrophoresis.

Agarose gel electrophoresis
Gel electrophoresis separates molecules according to size; the size range of the molecules separated depends upon the pore size of the gel. The pore size of agarose gels is a function of agarose concentration; thus, large pore sizes (low percent agarose, e.g. 0.2%) are useful for separating large DNA fragments (5–60 kb), and small pore sizes (higher percent agarose, e.g. 1.7%) for separating smaller fragments (0.1–2 kb). The migration of fragments within a gel is approximately inversely proportional to the logarithm of the molecular weight; to determine size, one generally runs a standard of known molecular weight and compares the sizes of bands of interest by measuring distance from the origin and plotting on semilog paper.

There are three commonly used buffer systems for electrophoresis of DNA. These are Tris acetate, Tris borate and Tris phosphate. The major differences among these relate to ionic strength and buffering capacity. Low ionic strength buffers permit more rapid movement of the DNA fragments through the agarose or acrylamide matrix. They have low conductivity and therefore are not so prone to overheating at higher voltages. They do not generally give bands as sharp as one can obtain with higher ionic strength buffers such as Tris borate and Tris phosphate. These latter have the additional advantage of higher buffering capacities; their disadvantage is a tendency to overheat. It is good practice in any case to recirculate the buffer and run the gels at low voltage gradients. The voltage drop is not critical for molecules in a size range below about 1000 bp (0.5 to 10 V/cm) but becomes very important for larger fragments; for these, voltage drops of 0.5 V per cm or less give much better resolution. The gel concentration is also important for large fragments. Fragments as large as 50 to

75 kb can be resolved at low voltage gradients in 0.2% agarose gels. At higher gel concentrations large fragments cannot be resolved even at very low voltage gradients and all run with the same mobility. This is presumably because the molecules 'snake' their way through the gel. One can be misled by running extracted DNA on the same gel with λ DNA and assuming that the unknown is the same size as similarly migrating λ DNA.

Care should be taken not to overload a gel (10–20 μg is an adequate amount for a 7 mm wide slot on a 10 × 14 cm gel) since this may cause bands to run in an anomalous fashion or may obscure other faint bands.

Agarose gel electrophoresis can be carried out using either horizontal or vertical gels, but the former have several advantages over the latter. First, with horizontal gels, it is possible to load 2 sets of samples by casting a second set of wells midway down the gel. Second, a horizontal gel tray can support the weak or low concentration gels which are useful for separating high molecular weight fragments. In addition, these submerged or 'submarine' horizontal gels are considerably easier to pour and handle than vertical gels.

The materials needed to carry out the gel electrophoresis protocol given here include the following:

1. electrophoresis apparatus, including a gel casting plate, a comb for forming wells, and an electrophoresis tank; IBI (International Biotechnologies, Inc.) and BRL (Bethesda Research Laboratories) both market these,
2. a power supply that can deliver constant voltages up to 100 V,
3. a UV transilluminator for visualizing the ethidium-bromide stained gel,
4. a camera for recording results,
5. agarose, gel running buffer (TEB 0.089 M Tris borate, 0.089 M boric acid, 0.001 M EDTA), gel loading buffer (see formula in previous section) and ethidium bromide.

Protocol

1. For separating DNA fragments in the range of 0.6–10 kb, we use an agarose gel concentration of 0.8%. To pour the gel: tape the ends of the gel casting plate with adhesive tape to form a tight seal. Insert the well-forming comb approximately 5 mm from the top of the gel tray. Approximately 100 ml of agarose solution is required for a medium-sized gel (10 × 14 cm). Thus, add 0.8 g agarose to 100 ml of 1 × TEB. The agarose is melted by heating to 100°C over a Bunsen burner or in a microwave oven. After swirling to insure even mixture of the agarose, ethidium bromide is added to a concentration of 0.5 μg/ml; the solution is allowed to cool to approximately 60°C and then poured into the gel casting tray. Pouring the agarose solution when it is too hot can cause the adhesive tape to weaken and the agarose solution to leak out. The gel is then allowed to harden by cooling to room temperature (20–30 minutes for a gel of this size).
2. After the gel has hardened, the tape is removed and the gel in its tray is submerged in the electrophoresis tank, which has been filled with 1 × TEB. The comb is gently removed, and the gel is now ready to be loaded.

3. To load the gel, samples are first mixed with 'loading' buffer (20% sucrose, 10% Ficoll, 100 mM EDTA, 1% bromophenol blue), the purpose of which is to increase the density of the samples to insure that they fall into the wells. (The bromophenol blue also allows one to follow the course of the electrophoresis; in a 0.8% gel, it migrates with fragments of 0.3 kb). Generally a 'marker' lane, containing fragments of known molecular weight, is also loaded. Samples can be loaded with a micropipet.
4. The gel is run at 1.5 V/cm at room temperature for approximately 16 hours. At the conclusion of electrophoresis, the gel is placed next to a ruler on a UV transilluminator and photographed.

Transfer of DNA

Following electrophoresis, the restricted DNA is transferred from the gel to a nitrocellulose (e.g. Millipore) or nylon membrane. It is the filter, to which the DNA has been transferred, that will be hybridized with the radiolabeled probe. Prior to transfer, it is necessary to denature the DNA by soaking the gel in alkali. This disrupts the hydrogen bonds in the double helix, producing two single stranded fragments. After the gel has been neutralized, the DNA is transferred to the membrane by passage of a high salt solution through the agarose gel. The rate of transfer of DNA from the gel to the membrane depends on the size of the fragments and on the porosity of the gel. In a 0.8% agarose gel, fragments less than 1 kb will transfer within one hour, whereas fragments greater than 15 kb will require an overnight transfer. In practice, transfers are usually left overnight. After transfer is complete, DNA fragments are permanently bound to the filter by baking.

A protocol for performing Southern transfer is given below (Fig. 2.4).

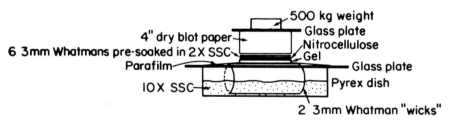

Fig. 2.4 Structure of a Southern transfer. See text for details.

1. Following electrophoresis, transfer the gel to a glass baking dish and rinse in several volumes of denaturing solution (1.5 M sodium chloride, 0.5 M sodium hydroxide) with gentle shaking at room temperature for 60 minutes.
2. Neutralize by soaking in 1 M Tris, pH 8, and 1.5 M sodium chloride at room temperature for 30 minutes. Change the neutralizing solution and continue soaking with gentle shaking for an additional 30 minutes. The pH should be less than 8.5.
3. In the next step, the transfer is set up as shown in Figure 2.4. First, fill a

pyrex dish with 10 × SSC and place a glass plate as a bridge across the top of the pyrex dish.
4. Pre-wet two 12 inch by 6 inch 3 mm Whatman wicks in 2 × SSC. These should extend from the reservoir of 10 × SSC and over the glass plate. With a glass rod or pipet, roll out bubbles between the wet Whatman papers and the glass plates.
5. Flip the gel over onto the Whatman paper and again roll out any air bubbles trapped between the paper and the gel. Mask the sides of the paper adjacent to the gel with parafilm. This prevents evaporation of SSC during transfer and forces the SSC to move through the gel.
6. Float a sheet of nitrocellulose cut to fit the gel on water until it is wet, then transfer to 2 × SSC for 5 minutes. Place the nitrocellulose onto the gel and roll out bubbles gently. Note that gloves should always be worn when handling nitrocellulose. If the membrane is contaminated by finger oils, it will never wet properly. If the nitrocellulose is not properly presoaked, transfer of DNA will be unreliable. Nitrocellulose is fragile and must be handled carefully especially after baking.
7. Cut six 3 mm Whatman papers to fit the gel and presoak these sheets in 2 × SSC. Apply these one at a time over the nitrocellulose, rolling out bubbles after each sheet is applied.
8. Cover the wet papers with 4 inches of dry blotting paper, either paper towels cut to fit the gel or S & S dot blot paper (Schleicher and Schull). On top of the dot blot papers, place a glass plate, and over the glass plate, a 500 g weight.
9. Transfer by incubating overnight at room temperature. Following transfer, the gel should be thin and translucent. All of the ethidium bromide fluorescence should be on the nitrocellulose, not the gel.
10. Peel off the gel and wash the nitrocellulose filter in 2 × SSC with gentle shaking for 5 minutes.
11. Air dry the filter for 10 minutes.
12. Place the filter between two dry Whatman papers and bake in a vacuum oven at 80°C for 2 hours. The baked filter may be stored in a sealed plastic bag at 4°C for several months, or pre-hybridized and hybridized immediately.

Hybridization

The fragments to be analyzed are detected by their hybridization to a single stranded radiolabeled probe. Methods for radiolabeling of fragments to be used as probes are described in Chapter 4 and Chapter 11 Part Two. In order to obtain interpretable results, it is important to prevent non-specific binding of the probe to the filter. Otherwise, the resulting areas of high background may obscure the bands of interest. In order to insure low background, filters are presoaked in a solution containing 5 × Denhardt's (1 × Denhardt's contains 1% each of Ficoll, bovine serum albumin [Miles Scientific, Pentax fraction] and polyvinylpyrrolidone) and heterologous DNA, which blocks further binding of DNA to the nitrocellulose. Following a several hour presoak or pre-hybridization, the radiolabeled probe is added. Factors affecting the rate of hybridization of the probe to its complementary strand on the filter include temperature, salt concentration, probe concentration, G-C content of the region of interest, and time allowed for the hybridization reaction.

The effects of these parameters on the melting temperature (T_m), the temperature at which the DNA 'melts' from the double stranded to the denatured state, are well understood. An increase in the salt concentration increases the T_m[10] and thus favors hybridization. Similarly, increases in the G-C content of the fragments of interest also confer a high thermal stability.[10] The optimum temperature for hybridization can be lowered by adding formamide to the hybridization solution; formamide reduces the melting temperature by 0.6°C for each 1% increase in formamide concentration.[11] We have had most success using a hybridization solution containing 6 × SSC at 65°C. Using higher salt concentrations promotes hybridization but also lowers specificity and increases background.

As a general rule, the smaller the volume of pre-hybridization and hybridization solutions, the better. This is for two reasons: first, the amount of radiolabeled probe needed is less; second, DNA/DNA hybridization occurs in a shorter period of time, since the concentrations of DNA are higher than in a larger volume. The time required for hybridization is a function of the DNA concentrations. The computation of kinetic constants in this situation is complicated by the fact that one of the DNA strands is immobilized, that is, is not free in solution. In the protocol given below, using a probe of specific activity, 2×10^8 cpm/μg, we would recommend a 16 hour hybridization with 50×10^6 cpm in order to detect a single copy gene in genomic DNA.

After hybridization, the unbound probe is removed with a high salt wash and mismatched hybrids are melted out with a low salt wash at 67°C. Filters are then autoradiographed for 48–72 hours. If desired, filters can be stripped of radiolabeled probes by washing under stringent conditions, and rehybridized to a second probe.

There are numerous protocols available for hybridization of Southern blots. The one given below has proven useful in our laboratories for detection of single copy genes.

1. The filter is pre-hybridized for 5 hours at 67°C in 6 × SSC, 10 mM EDTA, 100 mM potassium phosphate (pH 7) 5 × Denhardt's, 0.5% SDS, 0.25 mg/ml denatured salmon sperm DNA. A 10 ml mixture is sufficient for a medium sized blot (10 × 14 cm) and can be made up as follows: 3 ml 20 × SSC, 0.5 M EDTA (pH 8), 1 ml 1 M potassium phosphate (pH 7), 1 ml 50 × Denhardt's, 0.5 ml 10% SDS, 0.25 ml 10 mg/ml salmon sperm DNA stock, boiled, 4.05 ml of double distilled water.
2. Following pre-hybridization in a sealed plastic bag, the corner of the bag is clipped and the radiolabeled probe is added. We use 5.0×10^7 cpm of a nick translated probe with specific activity $1–2 \times 10^8$ cpm/μg. Prior to adding the probe, it is denatured by boiling for 5 minutes and cooling on ice for 2.5

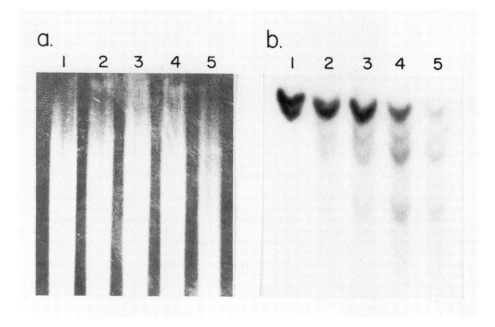

Fig. 2.5 (a) Ethidium bromide stained gel of genomic DNA prior to transfer to nitrocellulose. (b) Autoradiograph of nitrocellulose blot following hybridization to a radiolabeled probe.

minutes. The bag is resealed, placed at 67°C in a water bath and hybridization allowed to proceed for 16 hours.

3. Filters are washed, first in 2 × SSC, 0.5% SDS at room temperature (twice, for 10 minutes each), then in 0.1 × SSC, 0.5% SDS at 67°C (twice for 30 minutes each). Following the last low salt wash, the filters are air dried, wrapped in Saran wrap while still damp, and exposed at −70°C for 48–72 hours using Dupont lightning plus intensifying screens and Kodak XAR-5 X-ray film.

To strip filters for rehybridization to a second probe, the filters are incubated in 5 × Denhardt's and 0.5% SDS for 4 hours at 80°C. Filters can then be rehybridized as described above.

The most common technical problem encountered in Southern blotting is that of high background and/or low signal intensity. The most common causes of these problems are listed below.

1. Impurities in the agarose used to make the gel can result in high background. We have used both IBI (International Biotechnology, Inc.) and BRL (Bethesda Research Laboratories) agarose, but any high quality reagent grade preparation should be adequate.

2. Impurities in the Denhardt's solution used in the pre-hybridization-hybridization mix may also contribute to high background. It is important to use a high

quality grade of bovine serum albumin (e.g. Miles Scientific, Pentax fraction) in preparing the Denhardt's.

3. Contamination of radiolabeled probe with unincorporated radioactive nucleotide can also cause background problems, whether or not one uses column or precipitation methods to separate out unincorporated label (see Chapter 4 and Chapter 11 Part Two). For best results, the TCA precipitable counts should be at least 90% of the spot count.

REFERENCES

1 Blin N, Stafford D W. Isolation of high molecular weight DNA. Nucleic Acids Res 1976; 3: 2303–2308.
2 Gross-Bellard M, Oudet P, Chambon P. Isolation of high molecular weight DNA from mammalian cells. Eur J Biochem 1973; 36: 32–38.
3 Stafford D W, Bieber D. A rapid and efficient method for concentrating DNA. Biochim Biophys Acta 1975; 378: 18–21.
4 Maniatis T et al. Molecular cloning: a laboratory manual. Cold Spring Harbor Laboratory, 1982.
5 Kirby K S. Isolation and characterization of ribosomal ribonucleic acid. Biochem J 1965; 96: 266–269.
6 Kirby K S, Cook E A. Isolation of the deoxyribonucleic acid from mammalian tissues. Biochem J 1967; 104: 254–257.
7 Old J M, Higgs D R. Gene analysis. In: Weatherall, ed. Methods in hematology: the thalassemias. Edinburgh: Churchill Livingstone, 1983: p 78.
8 Fuchs, Blakesley. Guide to the use of type II restriction enzymes. In: Wu R, Grossman L, Moldave K, eds. Methods in enzymology. Vol 100. New York: Academic Press, 1983: p 3–38, 8.
9 Southern E M. Detection of specific sequences among DNA fragments separated by gel electrophoresis. J Mol Biol 1975; 98: 503–517.
10 Marmur J, Doty P. Determination of the base composition of deoxyribonucleic acid from its thermal denaturation temperature. J Mol Biol 1962; 5: 109–118.
11 Bluthmann H, Bruck D, Hubner L, Schoffski A. Reassociation of nucleic acids in solutions containing formamide. Biochem Biophys Res Commun 1973; 50: 91–97.

3
RNA isolation and processing
D. Leibowitz K. Young

INTRODUCTION

Handling RNA requires a combination of religious attention to detail and the approach of a surgeon to maintaining a sterile field during an operation. The eukaryotic cell contains four species of RNA, including messenger RNA (mRNA), ribosomal RNA (rRNA), transfer RNA (tRNA), and heterogeneous nuclear RNA (hnRNA). Laboratories investigating the expression of an individual gene are interested in analyzing mRNA, which represents roughly 5% of total cellular RNA. Ribosomal RNA accounts for about 80% of total cellular RNA, and tRNA and hnRNA account for the remainder. Messenger RNA is located primarily in the cytoplasm, with small amounts in the nucleus. Ribosomal and transfer RNA are located in the cytoplasm; hnRNA includes the precursors to mRNA and is located in the nucleus.

One of the first decisions to be made in analyzing the transcript from a gene is whether to analyze total cellular RNA or to purify mRNA. If the transcript from the gene is abundant it may be possible to work with total cellular RNA. If the gene's transcript is less common, it will be necessary to perform maneuvers which will remove RNA species other than mRNA. For a scarce transcript, a rough rule of thumb is that starting with 1000 μg of total cellular RNA from approximately 1×10^8 cells will provide enough concentrated mRNA (i.e. poly-A selected RNA) for analysis. If the transcript is abundant it may be possible to complete an analysis with as little as a few hundred μg of total cellular RNA. In some circumstances it will be necessary to separate the cytoplasmic and nuclear RNAs, which can be done relatively easily.

There are a number of differences between RNA and DNA which have large effects upon the way a worker must behave with them. The crucial difference is not inherent to the nucleic acid itself, but rather to the enzyme which degrades it. While DNase is a fragile enzyme, easily incapacitated by mild protein denaturants, heat, or physical trauma, RNase is a hardy enzyme which is extremely difficult to inactivate. A whole set of rigorous protocols are required to inactivate RNase, and it is impossible to be overly careful in protecting RNA from the ever present menace of RNase. Because RNA is so difficult to isolate without degradation, RNase is thought to exist on our very fingertips, always ready to digest the nucleic acid if we touch a non-gloved hand to a pipet tip or a test tube.

RNA is a single stranded nucleic acid, so that it will not be broken by physical shearing in the way that high molecular weight DNA is degraded. Since RNA has a hydroxyl moiety on the 2' carbon, the phosphodiester bond between adjacent bases is subject to being broken in alkali, so that moderately strong bases will destroy RNA. DNA is denatured into single stranded form by exposure to alkali, but it is not degraded.

RNA transcripts are commonly analyzed using either Northern blots, S_1 nuclease and RNase protection assays, or by primer extension. RNA is also isolated as the first step in making a cDNA library. This chapter will discuss the analysis of RNA with Northern blots, while some of the other procedures are discussed in other sections of the text.

Commandments
General rules to be observed when handling RNA include:
1. *No bare hands*, always wear gloves. They should be plastic gloves which are clean, but do not need to be sterile. An unavoidable article of faith will be that particular laboratory items (such as gloves) are free of RNase without specific treatment.
2. The traditional wisdom for rendering glassware free of RNase requires baking for at least 12 hours at 180°C. This would also apply to stir-bars, spatulas for weighing chemicals, and anything else which will be used in the processing of the nucleic acid. In practice, thorough autoclaving is usually accepted as adequate for decontaminating utensils.
3. Traditional wisdom has also held that water (deionized and distilled, of course), or other solutions, should be treated with 1% diethylpyrocarbonate (DEPC) to denature RNase. The DEPC is then removed by autoclaving the solution. Tris causes DEPC to break down, so DEPC cannot be added to Tris-containing solutions. Rather, Tris stocks should be made in DEPC-treated distilled water. Reagents to be used with RNA should be clearly labelled, and a separate stock should be maintained. These reagents should always be manipulated while wearing gloves, and nothing but baked or autoclaved utensils should be used with them.
4. Plasticware which is supplied as 'sterile' is assumed to be free of RNase. If desired, the plasticware may be rinsed with a 1% solution of DEPC in distilled water.
5. Solutions and reagents. Separate stocks should be maintained for working with RNA, and scrupulous technique must be observed: pour aliquots of chemicals out of containers, rather than sticking utensils into the chemical. When a utensil is used it should have been baked or autoclaved as in Rule 2.

PERTURBING THE CELL

As long as a cell is alive, its endogenous nucleases are contained within vesicles (lysosomes), protecting the cell's own nucleic acids from degradation. As soon as the cell dies, its previously captive RNase attacks the RNA molecules within the cell. Therefore, the first step in isolating RNA is to immerse the cell in a buffer which will inactivate the RNase long enough to permit the isolation of the nucleic acid. There are enormous differences in the amount of nuclease present in different types of cells. Lymphocyte cell lines are quite civilized, and it is possible to isolate intact RNA from them using a buffer that employs only SDS as a protein denaturant, along with EDTA, to inactivate the RNase. For polymorphonuclear granulocytes it is necessary to employ a very powerful denaturant to protect the RNA. One common procedure employs guanidinium thiocyanate (a.k.a. isothiocyanate), which is such a powerful denaturant that it is described as a 'chaotropic agent'. This procedure is well described in the Maniatis Manual[1] and in references[2] and[3] (see Protocol 1A at end of this chapter). A variant on this protocol uses the slightly less chaos-inducing guanidine hydrochloride.[2] A particularly convenient procedure, which is also highly protective, uses a LiCl/urea buffer for lysis of the cells (described below).[4,5]

After the cells have been lysed in an appropriate buffer, it is necessary to separate the RNA from the DNA and protein which are present. One very convenient and popular technique involves layering the cell lysate over a CsCl cushion and then spinning the sample in an ultracentrifuge.[3,6] The RNA will pellet (along with some CsCl), the DNA will be concentrated in a layer just at the top of the cushion, and protein will float to the top of the gradient. This allows an investigator to separate the protein away from both the RNA and the DNA of the cell, thus providing high molecular weight DNA in addition to RNA.

When the CsCl cushion procedure is used with cells that are low-risk (have very little RNase), the lysis buffer can be Tris/EDTA/sarcosyl (see Protocol 1 and Ch. 13), and the yield of DNA and RNA is excellent. The procedure becomes considerably less useful when working with high-risk cells or tissues (plentiful RNase) such as granulocytes, intestinal samples, and many solid tissues. With the high-risk tissue it is most common to use the guanidinium thiocyanate lysis buffer.[1,2] When the cells are lysed in the guanidinium reagent there is a less effective disruption of the cellular constituents. If the lysed cells are directly centrifuged over CsCl cushion the yield of RNA will be very low, a few per cent of the yield with the Tris/EDTA/sarcosyl. Also, the protein separates poorly from the DNA, so that although recovery of DNA is possible, it is not at all as clean as with the Tris/EDTA/sarcosyl procedure. In order to get an adequate yield with the guanidinium procedure it is necessary to homogenize the sample, preferably with a Polytron (Brinkmann, Westbury, N.Y.)*, in which case it will not be possible to recover any intact high molecular weight DNA. Even after good homogenization, the yield of RNA per cell from the guanidinium procedure is only about 10% of the yield from the Tris/EDTA/sarcosyl procedure.

An additional problem with RNA obtained by the guanidinium procedure is that it does not give crisp bands on Northern blots done with formaldehyde gels.

* Editor's note: an alternative device used to good effect in our laboratory is the Tekmar homogenizer.

If the RNA is to be analyzed by S_1 nuclease or RNase protection assays then the guanidinium protocol is perfectly adequate. If you choose to use the guanidinium protocol, and to centrifuge the lysate through a CsCl cushion, you must use polyallomer centrifuge tubes, because the guanidinium will destroy cellulose tubes.

The procedure used routinely in the author's laboratory is one using LiCl/urea.[4,5] The urea denatures protein, including RNase, while the LiCl precipitates the RNA. This protocol does require homogenization of the cells or tissue, so that it is not possible to isolate both high molecular weight DNA and intact RNA from a single sample. The yield is quite high, so it may be practical to consider splitting a sample into two portions, one of which is processed for RNA with LiCl/urea, while the other is processed for DNA by the most convenient method. The protocol for this procedure is given at the end of the chapter, and the procedure is discussed below. It should be noted that homogenization with a Polytron has consistently given a yield at least two-fold higher than the yield obtained using a propeller-blade type of homogenizer.

LiCl/urea

This procedure reproducibly works well for the buffy coat of peripheral blood, tissue culture cells, or solid tissue.

Solid tissue gives a good yield when it is rapidly cut into small pieces, immersed directly in the lysis buffer at 4°C, and homogenized. For tissue which has been frozen in liquid nitrogen in a single piece, it is necessary to first disrupt the tissue so that it is in small pieces prior to homogenization. This may be done either by processing the tissue in liquid nitrogen in a 'waring' blender, or by grinding it in a mortar and pestle on dry ice. In either case, once the tissue has been broken into small pieces, it can be added directly to the lysis buffer and homogenized.

The procedure provides protection from RNase which is acceptable for use with high-risk tissues, such as peripheral blood granulocytes or samples of small intestine. With peripheral blood the sample is first centrifuged for 20 minutes at 1800 rpm in an RNase-free tube, after which it is possible to remove and discard the plasma, and then remove the buffy coat. The buffy coat is transferred directly into a 250 ml Corning centrifuge tube, which is resting on ice. If the sample is very small (3 ml or less) then the volume of lysis buffer (see Protocol 1) used is typically 60 ml. Use 120 ml of lysis buffer for a sample which is larger than 3 ml. The quantitation of volumes by 60 ml fractions is pragmatically determined by the convenience of using 30 ml Corex tubes for the overnight incubation and the initial centrifugation on the second day of the procedure.

Since the lysis buffer and the LiCl/urea wash buffer are both stored at 4°C, the SDS in the lysis buffer is a precipitate. The lysis buffer should not be warmed up in an attempt to bring the SDS back into solution. The buffer should be swirled before use, so that the SDS is evenly distributed in the buffer, and then the suspension poured into the container being used for the homogenization. After the sample has been added to the large Corning tube the lysis buffer is swirled to distribute the precipitated SDS, and then poured into the Corning tube, up to the appropriate volume mark on the side.

A small amount of antifoam-A (Sigma, St. Louis, MO.) should be added to the lysate, and then the sample can be homogenized. The lower end of the Polytron's generator must always be immersed in liquid when homogenizing. Two sets of three 30 second bursts has been adequate for complete homogenization. Let the sample cool for a few minutes on ice between the first and second sets of homogenization. The 10 mm Polytron generator works well for volumes between 60 and 120 ml. If the sample volume is greater than 120 ml it will be necessary to divide the sample between two Corning tubes for homogenization. For a sample less than about 20 ml it may be possible to homogenize in a 50 ml plastic tube, but great care will be necessary to avoid splashing. For very small sample volumes it may be best to purchase the smaller 7 mm Polytron generator.

Following homogenization the RNA is precipitated by incubating the sample at 4°C overnight. We have not studied the stability of the sample if left at 4°C for longer than 24 hours. After this precipitation, the sample is centrifuged at 10 000 rpm, 4°C, for 30 minutes. The precipitate should be flat and solid in appearance; if the precipitate is thick and gelatinous the homogenization was inadequate and the sample will have a large amount of DNA in it. The supernatant is decanted, the sides of the tube are wiped with a Kimwipe, and 10 ml of the LiCl/urea wash buffer are added. The tubes are then vigorously vortexed: this usually generates a suspension, since the precipitate does not re-dissolve. After vortexing, the sample is spun for 15 minutes, at 10 000 rpm and 4°C, after which the supernatant is again decanted, and the sides of the tube are wiped very carefully, to remove as much of the LiCl/urea wash buffer as possible.

The pellet is then dissolved in a convenient volume of 100 mM Tris, pH 7.5/5 mM EDTA/0.2% SDS. This volume is generally 5 or 10 ml. It is usually necessary to warm the sample for a few minutes in a 37°C water bath, with intermittent vortexing, to get the pellet dissolved. The sample is then extracted with phenol/chloroform/isoamyl alcohol (50 : 48 : 2). Do not substitute a separate extraction with phenol, followed by another separate extraction with chloroform/isoamyl alcohol, since the poly-A tails of the mRNA may be trapped at the interface between phenol and the aqueous buffer. Separate the phases after vortexing by centrifuging for 10 minutes at 7000 rpm.

Save the organic phase, while the aqueous phase is removed and added to another tube which contains an equal volume of chloroform/isoamyl alcohol (24 : 1). Set the tube with the aqueous phase and the $CHCl_3$ aside for a moment (at room temperature), while proceeding to re-extract the organic phase. Add a volume of 100 mM Tris, pH 9, equal to the original aqueous volume, to the phenol/chloroform that was saved from the first extraction. Vortex, and then centrifuge for 10 minutes at 7000 rpm. Remove this aqueous phase and add it to the aqueous phase which was set aside with the $CHCl_3$; the volume of the aqueous phase will now be 2 × that of the organic phase. Vortex the tube with the sample and the chloroform, and then centrifuge for 10 minutes at 7000 rpm.

Remove the aqueous phase, add Na acetate to a final concentration of 0.3 M (0.1 × volume of 3 M Na acetate), followed by 2.2 volumes of EtOH, and incubate at −20°C. The RNA will be stable under EtOH for a period of time up to 6 months. If it is going to be stored for a long period of time before further processing a safer state for storage would be as an aqueous solution at −70° to −135°C. To do this, centrifuge the EtOH precipitate at 10 000 rpm for 30 minutes at −20°C, pour off the supernatant, rinse the pellet with 70–80% EtOH at −20°C, spin again for 20 minutes at 10 000 rpm, and again pour off the supernatant. Dry the inside of the tube well with a Kimwipe, dry the pellet under a vacuum, and resuspend it in small volume (0.5–1 ml) of 10 mM Tris, pH 7.5/1 mM EDTA/0.1% SDS. (Editor's note: if you anticipate using the RNA in enzymatic reactions, e.g. cDNA synthesis, omit SDS from storage buffers, or reprecipitate and resuspend in SDS free buffer prior to use.) This is a convenient time to measure the amount of RNA present by taking a small aliquot (5–10 μl) of the RNA solution to read the optical density (OD). (1 mg of RNA/ml = 25 A_{260} Units in a quartz cuvette with a 1 cm light path.) Place the sample in a secure plastic tube and store in an ultra-low temperature freezer or liquid nitrogen.

Regardless of the method chosen, the investigator will now have a sample of RNA in aqueous solution. If the transcript of interest is abundant, it will be possible to analyze the sample without further processing. Unfortunately, many transcripts are so scarce that it will be necessary to separate the mRNA away from the rest of the total cellular RNA for adequate analysis. The most commonly used procedure depends upon binding of the poly-A tail on the mRNA to oligo-dT as a way of enriching for the mRNA.

OLIGO-dT CELLULOSE CHROMATOGRAPHY

Messenger RNA has a long tail of adenylate residues (poly-A, or pA) at its 3′ end. This pA tail is unique to mRNA, and is the characteristic which is most generally used to enrich the mRNA content of a sample. Since there is a limit to the size of the bulk RNA sample which can be analyzed, whether the analytic procedure is blot hybridization or a nuclease protection assay, if a large amount of mRNA needs to be analyzed, then the non-mRNA has to be removed. Attaching an oligomer of thymidylate (oligo-dT) residues to a solid support provides a hook which can be used to snare the mRNA away from the other RNA species.

The solid support most commonly used is made of small particles of cellulose, to which the oligo-dT is added. The RNA sample is initially allowed to incubate with the oligo-dT cellulose in a high salt buffer, which stabilizes the hybridization between the pA tail and the oligo-dT. The mixture is then progressively washed with buffers of lower salt, until the final wash which contains no salt, in which the RNA with pA tails is released.

For most samples it is convenient to make a small column with the oligo-dT cellulose, and then to allow the RNA solution to slowly drip through the column. This protocol is well described in the Maniatis Manual.[1] It is quite important to allow the sample to return to room temperature after it is heated for denaturation. A higher temperature will destabilize the pA:oligo-dT hybridization (having the same effect as a low salt buffer) and will decrease the yield of the column.

The progress of the separation is followed by reading the OD of the eluate. The concentration of mRNA in the no-salt fractions is quite low, so in the author's laboratory the OD is measured by placing the sample itself into a cuvette: needless to say, the cuvette has to be extremely clean, since the RNA sample itself will be in contact with it. The cuvettes are generally prepared by washing with chromic acid cleaning solution, followed by washing with copious amounts of water. In practice, almost all of the polyadenylated RNA will be eluted in the first 3 ml of the no-salt wash. It is not uncommon to run the separated poly-A^+ RNA through an oligo-dT cellulose column for a second time in order to increase the enrichment for mRNA. The fractions with RNA should be pooled, brought to the appropriate salt concentration, and precipitated with EtOH for storage.

A rough estimate of yield is that about 3-4% of the total cellular RNA will be eluted in the final no-salt wash. The enrichment for mRNA is roughly 20-fold, though there is almost always a small amount of rRNA still present. It is also possible to process the RNA by mixing it with the oligo-dT cellulose as a slurry, then centrifuging the cellulose to the bottom of the incubation tube, followed by resuspension in the next aliquot of buffer. This sort of batch processing is less commonly used, and is not as reproducible as a column.

It is possible to purchase oligo-dT cellulose from a number of manufacturers, and with varying lengths of dT. For most purposes it is not necessary to have the extremely long lengths of dT that are most costly. Oligo-dT cellulose was originally manufactured by Collaborative Research (Lexington, MA; Type 3) and they continue to be a source of high quality material. The author's laboratory has most recently been using a Pharmacia (Piscataway, N.J.) product, Type 7, which has been quite dependable.

The columns can be reused a few times, but when the efficiency begins to drop they should be discarded. Save the eluates from the high and medium salt washes so that the RNA can be recovered in the event that some egregious error is discovered. Note that the eluate from the high salt wash should be diluted with distilled water to decrease the salt concentration prior to EtOH precipitation. We have not been successful in attempting to monitor degradation of the oligo-dT tails on the cellulose through the use of labeled pA.

ANALYZING RNA — NORTHERN BLOTTING

The most convenient procedure for the electrophoresis and blotting of RNA employs a formaldehyde/agarose gel. Transferring the nucleic acid out of the gel and onto a nitrocellulose membrane has been termed Northern blotting (as distinct from the blotting of DNA, which is termed Southern blotting (see Ch. 2) after Professor Ed Southern). The protocol for running RNA samples in formaldehyde/agarose gels, and transferring them to membranes (cf Ch. 2) is well described in the Maniatis Manual, the papers by Patricia Thomas,[7,8] and chapter 13. 1–10 µg of mRNA or 5–50 µg of total RNA are usually analyzed in each gel lane.

Though the above references suggest that running an RNA gel with ethidium bromide (EtBr) in the gel may interfere with transfer, many laboratories routinely run RNA gels with EtBr in the gel and do not have difficulty with transfer to membranes. When the transcript being investigated is present in a very low copy number it is most prudent to omit the EtBr, since even a slight decrease in the efficiency of transfer may make it impossible to detect the transcript. If the EtBr is omitted, it will be necessary to run another sample of RNA in an adjacent lane in order to determine the distance migrated by the 28S and 18S rRNAs. It is simple to run an extra lane with 10–20 µg of poly-A⁻ RNA as a marker at the same time that the other samples are run. The lane with the marker RNA can be cut off of the gel and stained with EtBr when the other part of the gel is being blotted. Incubating the marker lane in the running buffer with small amount of EtBr will make it possible to measure the distance migrated by the rRNA (if there is too much EtBr when viewed with UV light, just destain the gel in running buffer without EtBr).

The distance migrated by the rRNA bands will make it possible to calculate the size of bands later identified in Northern blots from the same gel. This is done by plotting the square root of the molecular weight against the log of mobility.[9] The 28S rRNA has a molecular weight of 1.584×10^6 daltons (one base pair = 330 daltons), with a square root of 1258.6. The 18S rRNA has a molecular weight of 5.94×10^5, with a square root of 770.7. By using semilog paper and plotting the distance moved by these marker bands on the log scale, against the square of their molecular weight, as given above, it will be possible to plot a line which can be used to calculate the molecular weight of bands read off of the Northern plot (an alternative may be RNA markers, which have recently been marketed by Bethesda Research Laboratories).

Northern blotting analysis of RNA can be accomplished essentially as described for Southern blotting of DNA (see Ch. 2), except as noted here. (Obviously, the RNA is not digested with restriction enzymes prior to electrophoresis.) 1–10 µg of poly-A⁺ mRNA or 20–50 µg of RNA are loaded into each lane. Nitrocellulose remains a dependable membrane for Northern blots, even though other membranes may be supplanting it for Southern blots. There are two reasons for this:

1. RNA transferred to nitrocellulose from formaldehyde/agarose gels is permanently bound, so that the blot can be rehybridized many times without loss of signal, and

2. nitrocellulose used for Northern blots does not become as brittle as when used for Southerns, presumably because the hybridizations are at a lower temperature.

Many different combinations of hybridization conditions and washes give excellent results. The two types of conditions used in this lab are:

Hybridization mix:
50% formamide
0.1% SDS
1 × Denhardt's reagent (100 × Denhardt's = 2% bovine serum albumin/2% Ficoll/2% polyvinylpyrrolidone)
2 × SSC (20 × SSC = 3 M NaCl/0.3 M Na citrate)
50 μg/ml salmon sperm DNA
(Alternatively, use 4 × SSC plus 20 mM sodium phosphate buffer, pH 6.8)[10]
Pre-hybridization is for 8–16 hours at 42°C. For hybridization 5–10 ng/ml of probe (roughly 2×10^6 counts/ml) are used, with at least 30 μl of hybridization fluid/cm^2 of blot.
After hybridization for 16–48 hours, washes (for high stringency hybridizations) are
The first wash must be at room temperature to remove the formamide, or else all of the probe will be washed off.
2 × SSC for 20 minutes at *room temperature*
2 × SSC for 10 minutes at 68°C
1 × SSC for 10 minutes at 68°C
0.2 × SSC for 10 minutes at 68°C
(Alternative washes after hybridization with 4 × SSC and Na phosphate buffer:

2 × SSC	20 minutes	room temperature
2 × SSC	30 minutes	50°C
1 × SSC	30 minutes	50°C)

Working with RNA is fun and rewarding, just keep in mind the maxim provided by a researcher who had made many cDNA libraries: If they're not laughing at you, you're not being careful enough.

Protocol 1
Lysis with Tris/EDTA/sarcosyl (for cells without RNase, such as lymphocyte cell lines).
Materials: *1.35 g/ml CsCl* — 20 g CsCl
14.5 ml 20 mM Tris, pH 7.6/10 mM EDTA/0.1% sarcosyl
0.5 ml 10% DEPC in EtOH
1.0 g/ml CsCl in 20 mM Tris, pH 7.6/10 mM EDTA/0.1% sarcosyl
20 mM Tris, pH 7.6/10 mM EDTA
20% sarcosyl
10 mM Tris, PH 7.6/1 mM EDTA/0.1% SDS
10% DEPC in EtOH
0.5% DEPC in H$_2$O (aqueous dilution of 10% DEPC in EtOH)
1. Pellet the cells.
   Rinse with phosphate buffered saline (PBS), and pellet.
   Tap the tube to loosen the cell pellet.
2. For each 1 ml cell pellet or less, add: 6.65 ml 20 mM Tris, pH 7.6/10 mM

EDTA, 350 µl 20% sarcosyl, swirl the tube quickly, then add 7 g CsCl. Place on a rocking mixer until the CsCl is in solution.
3. Pipet a 3 ml cushion of *1.35 g/ml CsCl* into a Beckman SW41 (or SW40) centrifuge tube (polyallomer tube, rinsed with 0.5% DEPC). Layer the cell lysate over the cushion, adjust the volume as required with the *1.0 g/ml CsCl* solution.
4. Centrifuge for 24–48 hours in an SW41 (or SW40), 30 000 rpm 12°C.
5. To take down the gradients: remove the soft crust of protein from the top of the tube. Pipet off the top 5–7 ml of solution, until reaching a viscous layer which contains the DNA. Use a clean pipet to remove the DNA and set it aside (see Step 8).
6. Pour off the remaining few ml of solution above the pellet (a CsCl pellet is usually visible, if there is no visible pellet the RNA is usually still there, so proceed with the processing). Cut off the top two-thirds of the centrifuge tube, dry the walls of the remaining portion, and resuspend the pellet in 1–2 ml of 10 mM Tris 7.6/mM EDTA/0.1% SDS. It may be necessary to heat the tube in order to dissolve the pellet.
7. Transfer the dissolved RNA to an RNase-free tube and precipitate with salt and EtOH (see text, and Protocol 2).
8. Dialyze the DNA with two changes of approximately 100 × volume of 10 mM Tris, pH 7.4/1 mM EDTA/0.1% SDS, for at least 24 hours, to remove CsCl, then phenol extract and process as desired.

Protocol 1A
Guanidinium thiocyanate extraction of RNA. This procedure is identical to Protocol 1, except as noted here.
1. Instead of sarcosyl buffer, use guanidinium thiocyanate buffer:
    94.4 g guanidinium thiocyanate
    1.47 g Na citrate
    1.4 ml β mercaptoethanol
    100.0 ml H_2O
    (Final volume 200 ml)
2. Use 7–10 ml of solution per g of tissue or per cc of packed cells.
3. Add cells or tissue to guanidinium solution and *immediately* disrupt with homogenizer (see LiCl/urea extraction Protocol and text).
4. For each 7 ml suspension use 4 ml CsCl cushion;
    CsCl 5.7 ml:
    95.7 g CsCl
    NaOAc 0.8 ml of 3 M (pH 6)
    to 100 ml w/H_2O
    Proceed as in Step 2 of this Protocol.
5. The guanidinium procedure shears DNA. Step 8 of Protocol 1 is not applicable.
N.B. Both guanidinium and CsCl solutions should be filtered before use.

Protocol 2
LiCl/urea RNA extraction (for cells with large amounts of RNase).
Materials: *LiCl buffer*— 3 M LiCl
                6 M urea
                0.05% SDS
                50 mM Na acetate pH 7.5
    LiCl wash— 3 M LiCl
                6 M urea
                50 mM Na acetate
    Tris/EDTA/SDS— 100 mM Tris, ph 8
                5 mM EDTA
                0.2% SDS
    100 mM Tris, pH 9
    Phenol/CHCl$_3$/Isoamyl alcohol (IAA) (50 : 48 : 2)
    CHCl$_3$/Isoamyl alcohol (24 : 1)

Preparation of *LiCl buffer*:
— Dissolve LiCl and Na acetate, in a volume smaller than the final volume of the solution (e.g. 400 ml if the final volume is 500 ml).
— Autoclave.
— After solution has cooled, add urea and dissolve.
— Bring volume to {(final volume) − (volume of 10% SDS to be added)}
— Sterile Filters, and transfer to storage container.
— Add appropriate volume of 10% SDS (SDS will precipitate when solution is cool).

Prepare *LiCl wash* in same manner, but omit SDS.

1. Place cells to be lysed in Corning 250 ml centrifuge tube (for peripheral white blood cells, centrifuge heparinized blood sample for 20 minutes at 1500–1800 rpm, then discard plasma, and remove buffy coat into Corning tube).
2. Add *LiCl buffer*: For cell pellet < 3 ml, add 60 ml of *LiCl buffer*
                    3–6 ml, add 120 ml *LiCl buffer*
   Swirl, let stand on ice for few moments.
3. Homogenize with Polytron, 30 seconds × 6.
4. Transfer to RNase-free 30 ml Corex tubes.
   Precipitate overnight at 4°C.
5. Centrifuge homogenate 30 minutes, 10 000 rpm, 4°C.
   Pour off and discard supernatant.
6. Add 10 ml *LiCl wash*, vortex, centrifuge 15 minutes, 10 000 rpm. Pour off supernatant, keep tubes upside down and wipe as much wash buffer as possible off the sides with a Kimwipe.
7. Dissolve pellet in 5 ml *Tris/EDTA/SDS* (will probably require warming to 37°C and vortexing; large samples will require a larger volume for resuspension).
8. Add equal volume (5 ml in this example) phenol/CHCl$_3$/IAA.
   Vortex.
   Centrifuge 10 minutes, 7000 rpm, 4°C.

48   MOLECULAR GENETICS

9. Remove aqueous layer to separate tube with an equal volume (5 ml) of $CHCl_3/IAA$, and set aside at room temperature. Add equal volume (5 ml) 100 mM Tris, pH 9, to the phenol/$CHCl_3$.
   Vortex.
   Centrifuge 10 minutes, 7000 rpm, 4°C.
10. Remove the 100 mM Tris, pH 9, and add to the aqueous layer which was set aside (the aqueous layer is now 2 × the volume of the organic layer).
    Vortex.
    Centrifuge 10 minutes, 7000 rpm, 4°C.
11. Remove the aqueous layer, add Na acetate to 0.3 M, and EtOH precipitate.

Protocol 3
Formaldehyde agarose gel electrophoresis.
1. *10 × Gel buffer*
   0.2 M MOPS (41.8 g/l)
   0.05 M Sodium acetate (4.1 g/l)
   0.01 M Disodium EDTA (10 ml of 0.5 M)
   pH to 7.0
   Autoclave. May or may not change to yellow colour.
2. *Gel 1.8% agarose*
   72 ml water
   1.8 g agarose
   Boil to dissolve
   Add: 18 ml 37% formaldehyde
        10 ml 10 × gel buffer
   Pour the gel in the hood.
3. Running buffer
   Dilute 10 × gel buffer to 1 × Run a submarine gel.
4. Sample preparation
   Spin RNA out of ethanol in microcentrifuge. Speed-vac dry. Use 10–20 μg total RNA or 3–5 μg poly-A^+. Bring RNA up in 5 μl sterile water.
   Add: 10 μl formamide (Fluka, deionized) — 50%
        3.4 μl formaldehyde (37%) — 6.3%
   Heat to 60°C for 10 minutes.
5. *Sample buffer*
   50% glycerol
   1 mM EDTA
   0.4% bromophenol blue
   Add 4λ sample buffer to RNA prior to loading on the gel.
6. Electrophoresis
   Run 3 hours at 100 V or until blue dye has migrated to the bottom.
7. Blot as usual in 20 × SSC O/N. Bake blot 2 hours in vacuum oven.

Protocol 4
mRNA purificaton by oligo-dT cellulose chromatography.
1. Suspend 1 g of oligo-dT cellulose in about 20 ml sterile distilled deionized DEPC treated water in a 50 ml graduate or centrifuge tube. Disperse and suspend by *gentle* stirring. Allow resin to settle for about 5 minutes. 1 g of oligo-dT can bind 1 mg of mRNA.
2. Decant or aspirate supernatant. Repeat procedure 5–7 times. This removes 'fines' that cause oligo-dT cellulose columns to clog, causing very slow flow.
3. Pour de-fined cellulose into 5 ml syringe or chromatography column plugged at bottom with siliconized glass wool.
4. When resin has settled, wash with 30–40 ml high salt buffer (0.5 M NaCl, 10 mM Tris HCl, pH 7.5, 1 mM EDTA, 0.2% SDS). Check optical density of eluate at 260 nm to establish 'baseline'.
5. Dissolve mRNA in minimal volume (1–2 ml) of 10 mM Tris HCl, pH 7.5, 1 mM EDTA. Add equal volume 2 × high salt buffer (twice the concentration shown in Step 4). Heat sample at 60°C for 5–10 minutes. Cool in ice water to room temperature.
6. Pipet RNA sample onto column. Allow sample (up to 2.5 ml) to flow into the column. Clamp for 5–10 minutes to allow mRNA ample time to bind column.
7. If sample is in a larger volume, repeat Step 6 until entire sample has equilibrated with column. Collect eluate in a large (50 ml) sterile tube.
8. Pipet high salt buffer onto column and allow buffer to flow through by gravity (do *not* pump buffer!). Collect eluate in same tube until about 15 ml has passed through. (At this point, we add 2 volumes 95% EtOH to the eluate and save it at −20°C, just in case the column failed to retain the mRNA.)
9. After 15 ml high salt buffer have washed through column, begin collecting 3 ml fractions. Check optical density at 260 nm. Collect fractions until A_{260} is at baseline.
10. Let buffer layer that is above packed resin flow through until there is essentially no buffer layer above top of packed resin bed.
11. Add low salt buffer (10 mM Tris HCl, pH 7.5, 1 mM EDTA, 0.2% SDS) 1 ml at a time, allowing each aliquot to flow in before adding the next. Collect eluate in a 15 ml sterile tube (3 ml fractions).
12. Measure A_{260} of each fraction, in *sterile* cuvette. (See text.) Collect tubes with peak A_{260}s. The mRNA is almost always in the first 3–5 ml.
13. For very pure mRNA (e.g. for cDNA library construction), the mRNA is often repurified by a 'second pass'. Add an equal volume of 2 × high salt buffer to the mRNA fractions and repeat Steps 4–11. (Smaller wash volumes can be used; approximately $\frac{1}{2}$ the volumes shown.)
14. Add $\frac{1}{10}$ volume 3 M sodium acetate, pH 5.4, and 2.5 volumes 90% EtOH. Store at −20°C at least 6 hours. At this point the mRNA can be handled like other RNA samples described in the text.

REFERENCES

1. Maniatis T, Fritsch E F, Sambrook J. Molecular cloning: a laboratory manual. New York: Cold Spring Harbor Laboratory, 1982.
2. Chirgwin J M, Przybyla A E, MacDonald R J, Rutter W J. Isolation of biologically active ribonucleic acid from sources enriched in ribonuclease. Biochemistry 1979; 18: 5294–5299.
3. Glisin V, Crkvenjakov R, Byus C. Ribonucleic acid isolated by cesium chloride centrifugation. Biochemistry 1974; 13: 2633–2637.
4. Auffray C, Rougeon F. Purification of mouse immunoglobulin heavy-chain messenger RNAs from total myeloma tumor RNA. Eur. J. Biochem 1980; 107: 303–314.
5. Leibowitz D, Cubbon R, Bank A. Increased expression of a novel *c-abl* related RNA in K562 cells. Blood 1985; 65: 526–529.
6. Young K, Donovan-Peluso M, Bloom K, Allan M, Paul J, Bank A. Stable transfer and expression of exogenous human globin genes in human erytholeukemia (K562) cells. Proc Natl Acad Sci USA 1984; 81: 5315–5319.
7. Thomas P S. Hybridization of denatured RNA transferred or dotted to nitrocellulose paper. Methods Enzymol 1983; 100: 255–266.
8. Thomas P S. Hybridization of denatured RNA and small DNA fragments transferred to nitrocellulose. Proc Natl Acad Sci USA 1980; 77: 5201–5205.
9. Lehrach H, Diamond D, Wozney J M, Boedtker H. RNA molecular weight determinations by gel electrophoresis under denaturing conditions, a critical reexamination. Biochemistry 1977; 16: 4743–4751.
10. Anderson M, Young B M. In: Hames B D, Higgins S J. Nucleic acid hybridization: a practical approach. Oxford: IRL Press, 1985: p 95–98.

4
Hybridization probes in molecular hematology

T. A. Rado   L. Ratner   J. M. Downing

The ability to find a specific DNA or RNA sequence in an extremely complex population of sequences depends on the availability of hybridization probes. Three major classes of probe are currently available to the molecular biologist. These include restriction fragments which contain the region of interest, oligonucleotide probes and, most recently, *in vitro* synthesized RNA probes.

The most commonly used probe is characterized by a relatively large piece of DNA, often recovered by endonucleolytic excision (restriction endonuclease digestion) from a plasmid or bacteriophage into which it has previously been cloned. The actual sequence of the piece may or may not be known.

Two major types of oligonucleotide probe have found common use. One is synthesized from the known sequence of a gene or DNA region, while the other is designed on the basis of amino acid sequence data. The important difference between these, of course, is that probes of the latter variety are by definition degenerate, and constitute families of oligonucleotides. The magnitude of these families depends on the length of the probe and the level of codon degeneracy defining the amino acids comprising the region from which it was synthesized.[1] The actual complexity of a family of oligonucleotide probes can sometimes be reduced by choosing codons on the basis of known codon preference in the organism being studied, and by compromising base specificity in the third, or 'wobble' position of the codon.[1,2]

The newest (and perhaps most powerful) tools for finding sequences in complex nucleic acid blots are RNA probes. These are invariably synthesized from plasmids bearing a subcloned DNA region and the promoter/primer for a DNA-dependent RNA polymerase.

In this chapter, all three types of probes will be discussed, and the authors will provide the protocols they use for efficient sequence detection. Recently, considerable interest has been attracted by the substitution of non-radioactive labels for ^{32}P-labeled nucleotides. These technologies will not be discussed in this section, and the reader is referred to recent articles in the field for further information.[3-5].

MODERATE LENGTH DNA PROBES AND LABELING OF DNA FRAGMENTS

Double stranded DNA fragments may, depending upon the use for which they are intended, be labeled randomly throughout their length, or specifically at the 3' or 5' end. The classical method for obtaining high specific activity probes has been through a technique known as 'nick translation'. This approach depends upon the introduction of single strand nicks into DNA with extremely dilute DNase I, followed by treatment with DNA polymerase I in the presence of excess nucleoside triphosphates, one or more of which carries a radioactive phosphate in the α-position.[6,7]

A number of protocols for this procedure have been published, and kits containing buffers, dNTPs and even radionuclide are widely available. The reaction depends upon the polymerase I-catalyzed exonucleolytic removal of bases from the double helix in a 5' to 3' direction, accompanied by the simultaneous addition of bases in the 3' to 5' direction.

The intrinsic inefficiency of this approach, the relatively large amounts of isotope needed to obtain high specific activities, and the propensity of the enzyme to create internal hairpin loops[6] has resulted in a continuing search for improved methods of DNA labeling. One such approach, which has been used successfully in our laboratory, depends upon the kinetics displayed by the dual actions of the enzyme T_4 DNA polymerase. Advantage is taken of the dNTP-dependent rates of the 3' to 5' exonuclease and the 5' to 3' polymerase activities of the enzyme.

In the absence of dNTPs the enzyme will successively remove bases from the 3' ends of a double stranded DNA regardless of the configuration of these ends (blunt, 3'-extended, or 5'-extended). The rate of exonucleolytic activity is independent of sequence, and is linear with time. Under standard conditions (see Fig. 4.1) bases are removed from the 3' ends at a rate of approximately 40 per minute.[8,9]

Addition of dNTPs to the reaction mixture inhibits the exonuclease reaction and permits the polymerase activity of the enzyme to resynthesize DNA in the 5' to 3' direction. Thus, the exonucleolytically treated DNA acts as a primed template for the polymerase, and the end result is a perfectly blunt-ended double stranded molecule. Labeling is accomplished by replacing one or more of the cold dNTPs with the α-^{32}P-labeled nucleoside triphosphates. Using a single nucleoside triphosphate as label (dCTP, α-^{32}P, 3000 Ci/mmol) we routinely obtain activities as high as 10^9 cpm/μg DNA. Activities as great as 10^{10} cpm/μg have been reported when the reaction is carried out in the presence of all four labeled dNTPs.

The labeling methods described thus far, nick translation and replacement synthesis with T_4 DNA polymerase, require relatively large amounts of radioactive nucleoside triphosphate. The polymerization step of the nick translation reaction is linearly dependent upon dNTP concentration, and in the presence of an excess of three 'cold' dNTPs to drive the reaction, the concentration of labeled nucleoside is rate limiting. A similar situation holds for the T_4 polymerase reaction.

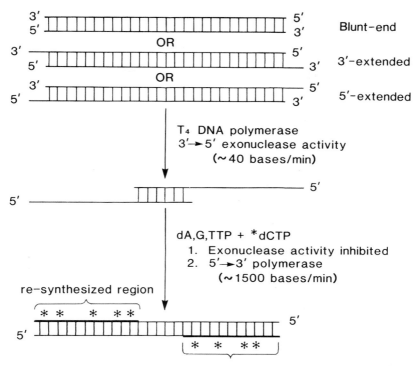

Fig. 4.1 Fragment labeling with T_4 DNA polymerase. The reaction is allowed to proceed in two stages. The following solutions are required: $10 \times$ TP buffer (0.5 M Tris acetate, pH 8.5, 0.5 M KAc, 50 mM each dATP, dTTP, and dGTP), α-[^{32}P]dCTP (3000 Ci/mMol). The first stage of the reaction contains 1 μl $10 \times$ TP buffer, 1 pmol DNA fragment, 1 unit T_4 DNA polymerase and water to 10 μl. The mixture is incubated at 37°C for the time required (calculating from the exonucleolytic rate of 40 bases/minute). Then 10 μl (250 μCi) dCTP, 2 μl dNTP $\times$ 3 mix, and 1 μl $10 \times$ TP buffer are added. The reaction is incubated for 15 minutes at 37°C. To prevent the reaction from reversing when the rate-limiting (labeled) dNTP is exhausted, 1 μl of 'cold' 1.0 mM dCTP is added at the end and incubation is continued for an additional 15 minutes. The reaction is stopped by the addition of 1 μl 0.25 M EDTA. We routinely remove unincorporated dNTPs by chromatography over a 1 ml column of G-50 Sephadex.

Recently a method for obtaining labeled DNA has been introduced which requires less radioactive triphosphate, and which has the additional advantage of functioning efficiently on DNA fragments embedded in agarose.[10,11] This technique takes advantage of the ability of oligonucleotides to prime DNA synthesis by the large (Klenow) fragment of DNA polymerase I. The DNA to be labeled is denatured by heat and then hybridized to a mixture of random hexanucleotides containing all possible 6-mer permutations of the four bases. These hybridized oligonucleotides prime the synthesis of DNA and, in the process, the incorporation of radioactively labeled dNTP. Incorporation of label by this method can reach efficiencies of 80–90%.

In our hands we have found this method to be excellent for labeling restriction fragments ranging in size from about 0.4 to 3 kb. Covalently closed circles of double stranded DNA must be opened with a restriction endonuclease prior to labeling, since denaturation and hybridization cannot occur in this configuration. Our protocol for preparing radioactive probe by the 'oligo-labeling' method is shown in Figure 4.2.

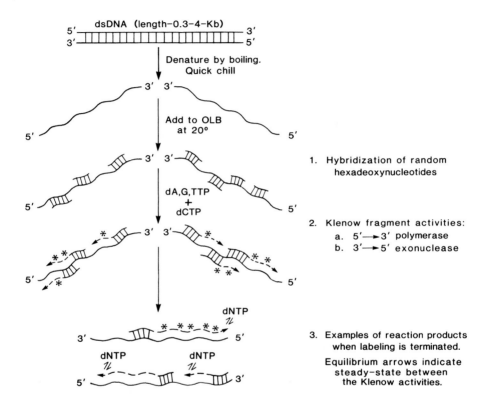

Fig. 4.2a Schematic representation of probe preparation by the 'oligo-labeling' method.[10,11] We prepare three solutions (each of which can be stored at $-20°C$) prior to the experiment. Buffer A is a mixture of dATP, dGTP, and dTTP (0.5 mM each) in 1.25 M Tris HCl, pH 8.0, 125 mM $MgCl_2$ and 250 mM 2-mercaptoethanol. Buffer B is 2.0 M HEPES titrated to pH 6.6 with 4.0 M NaOH, and Buffer C is a solution of random hexadeoxynucleotides (Pharmacia-PL or International Biotechnologies) in 10 mM Tris HCl, pH 7.4, and 1 mM EDTA. The three solutions are mixed at a ratio of A : B : C = 100 : 250 : 200. This is 'OLB Buffer'. The probe is denatured by boiling for 3 minutes, quickly chilled on ice immediately before use, and added directly to the reaction mix. The reaction mixture (final volume, 50 μl) consists of 10 μl OLB, 2 μl DNase-free BSA (10 mg/ml), denatured probe, 5 μl α-[^{32}P]dCTP (> 3000 Ci/mmol) 2 units of Klenow fragment (Boehringer-Mannheim) and water to volume. After overnight incubation at room temperature, we quench the reaction by adding 200 μl STOP solution (20 mM NaCl, 20 mM Tris HCl, pH 7.5, 2 mM EDTA, 0.25% SDS, 1.0 μM 'cold' dCTP). We have found that cleaner blots are routinely obtained if the probe is purified by passage over a column such as NENsorb (Dupont/New England Nuclear) or Elutip-D (Schleicher and Schuell) at this point. This method, without further modification, yields high specific activities for probe concentrations ranging from 10 ng to > 1 μg.

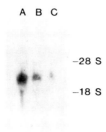

Fig. 4.2b Northern blot of total cellular RNA from the human myeloid cell line PLB-985 hybridized with an oligo-labeled human c-myc probe. Each lane contains 20 μg RNA. Lane A, uninduced cells; Lane B, cells induced with DMSO to differentiate in a granulocytic direction; Lane C, cells induced with phorbol myristate acetate (TPA) to differentiate in a monocytic direction (Tucker K, Rado T A, unpublished data).

Synthetic oligonucleotides

The development of methods for synthesis and use of oligonucleotides allows the detection and alteration of specific sequences within genes or transcripts. Here we will emphasize aspects and present examples of this technology which illustrate its usefulness in approaching research problems in hematology. Detailed protocols for the use of oligonucleotides as hybridization probes and for *in vitro* mutagenesis are provided in Chapter 11, Parts One and Two, respectively.

The synthesis of oligonucleotides by either the phosphotriester or phosphite-triester methods has been described in a number of reviews.[12–16] With current technology, the efficiency of synthesis is 99% per cycle, or 80% for an oligonucleotide 20 bases in length (a 20-mer). Further purification, if necessary, may be achieved by high pressure liquid chromatography and/or polyacrylamide gel electrophoresis. Sequences may be confirmed by partial chemical cleavage methods (Maxam-Gilbert sequencing).[16,17] The availability of automated synthesis at decreased costs now provides reagents for a wide range of studies. Most universities and companies will prepare oligonucleotides of the desired length and sequence for a modest fee.

Oligonucleotide probes may be labeled at their 5' or 3'-termini and used as hybridization probes. Labeling 5' termini with γ-[^{32}P] ATP (5000 Ci/mmol, using 5 μCi/pmol oligonucleotide) and T$_4$ polynucleotide kinase (10 units/10 pmol oligonucleotide, in a 50 μl reaction mixture) allows synthesis of probes with high specific activity. Under these conditions about 70% incorporation of radionuclide is achieved, and further purification of the probe is generally unnecessary.[18] Alternatively 3'-termini may be labeled with α-[^{32}P]dideoxy-ATP(7000 Ci/mmol) and terminal deoxynucleotidyl transferase.[19] The use of dideoxy-ATP in this reaction is less costly and achieves higher specific activities than were obtainable with an earlier method which depended on the incorporation of α-[^{32}P]3'-deoxyadenoside-5'-triphosphate (cordycepin triphosphate). This reagent is, however, still commercially available, and may be used for the labeling of 3' ends of single stranded oligonucleotides, and the 3' protruding ends produced by some restriction endonuclesses such as Hha I.[20]

The synthesis of double stranded molecules can be accomplished when short hybridization probes of higher specific activity are desired.[21] First the synthetic oligonucleotide is tailed at the 3' end with 15–20 adenylate residues using terminal deoxynucleotidyl transferase. The length of this tail is critical and must be calculated from incorporation data.[22] Then, oligo-dT$^{12-18}$ is hybridized and the complementary sequence synthesized in the presence of one or more α-labeled nucleoside triphosphates and the large (Klenow) fragment of E. coli DNA polymerase I. The product of this reaction may then be used as a double stranded DNA probe, or the labeled strand isolated on a denaturing polyacrylamide gel.[23]

Data exist, however, suggesting that the synthesis of second strand as a means of obtaining high specific activity probe may be unnecessary. Collins & Hunsaker[24] compared the allele specificity of oligonucleotides tailed at the 3' end by the addition of ~50 molecules α-[^{32}P]dATP with the same oligonucleotides labeled in the standard manner with a single atom of ^{32}P at the 5' end. They found that not only was the specific activity of the tailed molecule predictably greater than that of the 5' labeled probe, but also that there was no difference in the thermal dissociation of the two types of probes from the DNAs to which they were hybridized.

Oligonucleotide probes may be used for the detection of specific DNA or RNA fragments on nitrocellulose or nylon membranes. Precise conditions should be estsblished for salt and DNA concentrations and hybridization and wash temperatures, such that the signals for specific hybridization are strong, while non-specific hybridization is minimized. We have found a simple approach is to pre-hybridize filters in 5 × SSPE (0.75 M NaCl, 43 mM NaH$_2$PO$_4$, 4.4 mM EDTA, pH 7.4) containing 1.0% (w/v) sarcosyl and 100 μg/ml denatured salmon sperm DNA or yeast tRNA at 20–30°C. Hybridization to labeled probe is carried out in the same solution containing at least 10^6 cpm/ml of radioactive oligonucleotide. After hybridization, excess probe is removed and the membranes sre washed with 5 × SSPE, 1% sarcosyl four times for 10–15 minutes each. The

Fig. 4.3 (opposite) Demonstration of a deletion in proviral DNA with an oligonucleotide probe. DNA cloned in either a plasmid or bacteriophage from HTLV-Ib (pMCl^1, λEL-2, λEL-3) or HTLV-I (λ23-3, λSt) infected cells was digested to completion with Pst 1, Pst 1 and Cla I, or Pst 1 and Sst I as indicated. Samples were electrophoresed on a 0.8% agarose gel and transferred to nitrocellulose.[9] A 1.4 kb Pst I fragment of pMCl^1, including the majority of the pX region, was nick translated and used as a probe.[9] The filter was boiled and rehybridized with the oligonucleotide probe whose sequence is complementary to the 11 nucleotides deleted from λMC-1 and 2 flanking nucleotides on either side, as indicated at the bottom of the figure. The probe was labeled at the 5' end with T$_4$ polynucleotide kinase and the 3' end with terminal deoxynucleotidyl transferase. Hybridization was performed with 1.4 × 10^7 cpm of probe in 4 × SSC (0.6 M NaCl, 60 mM sodium citrate, pH 7.0), 4 × Denhardt's medium (0.08% polyvinylpyrrolidone, 0.08% bovine serum albumin, 0.08% Ficoll), 100 μg/ml salmon sperm DNA, 10 mM EDTA, 0.5% (w/v) sodium dodecyl sulfate (SDS), 10% (w/v) dextran sulfate at 20°C for 16 hours. The filter was washed with 4 × SSC, 0.02% (w/v) SDS, four times for 15 minutes each at 37°C and exposed to X-ray film for 30 minutes. The size of DNA fragments is indicated to the right of the figure.

a) pX probe

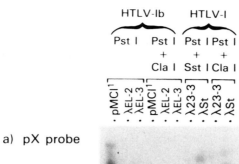

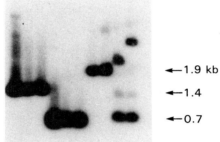

b) D17 probe

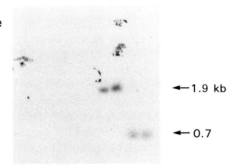

membranes are then wrapped in plastic wrap and exposed to film overnight without being allowed to dry completely. Successive washes at increasing temperature may then be performed until the signal-to-background ratio is optimized. (See Ch. 11 Part Two for detailed protocols.)

Figure 4.3 illustrates the use of a labeled 15 nucleotide oligomer corresponding in sequence to the 11 nucleotides deleted and the flanking two nucleotides present on either side from the pX region of a variant human T-lymphotropic virus, type I (HTLV-I), designated HTLV-Ib.[25] This probe, D17, failed to detect sequences in three different proviral clones derived from transformed T lymphocytes of the same patient. In contrast, a larger DNA probe, designated 'pX probe' readily detects sequences from this region of the viral genome. The same probe was then used to hybridize poly-A^+ RNA on a Northern blot. Transcripts of this portion of the pX region are detected in RNA from HTLV-I transformed cells, but not in RNA from cells infected with HTLV-Ib (Fig. 4.4). Transcripts from other regions of the viral genome are present in all three RNA preparations (not shown). This confirms that the region deleted in proviral copies of HTLV-Ib is not being complemented by other proviruses or cellular sequences, and thus this portion of the pX region is not required for maintenance of the transformed state.

The short length of oligonucleotide probes allows the discrimination of sequences which differ by only one nucleotide. Thus, one can theoretically use an oligonucleotide probe to detect point mutations by differential hybridization to Southern blots of genomic DNA. These methods have been applied successfully for the detection of globin and α_1-antitrypsin gene mutations in prenatal diagnosis or in the determination of heterozygosity.[26-29]

Oligonucleotides may also be useful for mapping exon-intron boundaries. An oligonucleotide probe complementary to the last 7-9 nucleotides of one exon and the first 7-9 nucleotides of the next exon will hybridize only with properly spliced mRNA species under appropriately stringent conditions.

Perhaps the most frequent application of oligonucleotide probes is in screening recombinant phage or plasmid libraries. If one has limited amino acid sequence data, an oligonucleotide or a mixture of oligonucleotides may be synthesized which is complementary to the sequence predicted to encode a portion of the protein whose cDNA gene is being sought. To limit artifactual hybridization to related but not identical sequences, the oligonucleotide should be at least 15 nucleotides in length, and regions of the amino acid sequence with the least codon degeneracy (e.g. methionine and tryptophan containing regions) should be chosen. Furthermore, one can limit the number of species in a mixed oligonucleotide probe by excluding those codons which are known to be utilized rarely in a given organism.[30,31] Where no obvious codon preference exists, G-T mismatches may be chosen in preference to A-C mismatches, to increase the stability of hybridization. (See Ch. 11 Part Two for detailed description and protocols.)

The successful detection of a single copy gene on a genomic Southern blot or a rare clone in a complex library by means of hybridization to a mixed oligonucleotide probe depends on several factors, some of which can be optimized. Since the dissociation of probe from immobilized nucleic acid depends on the

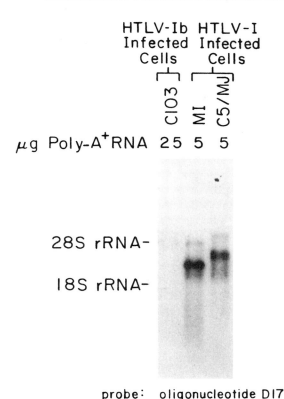

Fig. 4.4 Transcripts including the beginning of pX-I are expressed in HTLV-I but not HTLV-Ib infected cells. Poly-A$^+$ RNA was prepared from HTLV-Ib (C103) and HTLV-I (MI, C5/MJ) transformed cells and electrophoresed on a 1.0% formaldehyde agarose gel.[9] RNA was electrophoretically transferred to Zeta-probe® filters. The filter was hybridized with 1.4 × 10^7 cpm of the D17 oligonucleotide probe (see Fig. 4.1) labeled at both the 5' and 3' termini. Hybridization was performed as described in Figure 4.1, and the filters washed in 4 × SSC, 0.02% SDS at 20°C, 4 times for 15 minutes each. Autoradiographic exposure with an intensifier screen was for 4 hours. Positions of 28S and 18S rRNA markers are shown to the left of the figure.

melting temperature of the hybrid (T_d), the determination of maximally stringent conditions which still permit hybridization would be quite useful. In buffers such as SSC or SSPE, T_d at any given ionic strength depends in part on probe length, in part on the degree of mismatch, and in part on the G+C content of the ultimately 'correct' member of the probe mixture.

A method has recently been described in which the standard washing buffers are replaced by tetramethylammonium chloride (TMCL). This reagent eliminates the difference in melting temperature between A-T and G-C base pairs. Thus, at any given temperature, hybridization is solely dependent upon probe length and the perfection of match.[32] For relatively short probes (10–40 bases), careful control of wash temperature is all that is required to provide predictable levels of stringency. In our hands, at 64°C only, a perfectly matching 24-mer hybridized. At lower temperatures non-specific hybridizations were noted, and at 68°C all probe was removed.

A variety of applications utilize oligonucleotide primers for the synthesis of complementary DNA strands by the Klenow fragment of *E. coli* DNA polymerase I or reverse transcriptase. This methodology can be applied to dideoxy sequencing of DNA fragments in single or double stranded phage or plasmid preparations,[33-37] and even directly to cellular RNA.[38]

Oligonucleotide primers hybridized to poly A-containing RNA may be extended with reverse transcriptase to synthesize cDNA species for cloning or for the study of RNA initiation and processing. Mixed oligonucleotide preparations designed on the basis of amino acid sequence information have been successfully used as primers for cDNA synthesis (see reference[39] for review; also[40]). When sequence data on a gene is available, full length cDNA can be synthesized by using a 3' complementary primer to initiate first strand synthesis, followed by the priming of second strand synthesis with an oligonucleotide complementary to the 5' end.[41]

If the RNA in which one is interested has an unusually long 3' untranslated region or extremely low abundance, the likelihood of obtaining full length cDNA after oligo-dT priming is small. The human estrogen receptor, a low abundance protein, has recently been cloned using a mixture of every possible 13-mer as reverse transcriptase primer, and sucrose gradient enriched mRNA as template.[42] The use of random oligonucleotide mixtures to prime reverse transcription from an origin upstream from the poly-A region is a potentially powerful approach which will undoubtedly see wider application in the future.

New applications of oligonucleotides have been described and applied to the amplication and mapping of genomic DNA sequences from individuals suspected of having a hemoglobinopathy.[43] This method is called the polymerase chain reaction. Using these techniques, genomic sequences may be amplified from small quantities of tissue containing as little as 1 ng of DNA. As shown in Figure 4.5, two oligonucleotides are required. One is *complementary* to the 3' end, and one *identical* to the 5' end of the region of interest. These oligonucleotides are hybridized to genomic DNA and the chains extended with Klenow enzyme. The reaction product is denatured, and the cycle of hybridization, chain extension, and denaturation repeated. The entire process may be repeated up to twenty times in a single day, producing (in the study described by Saiki et al[43]) a 220 000-fold amplification of a 103 base pair region. The efficiency of each cycle in this application was 70–100%, but may be reduced if amplification of longer regions is attempted. Thus far, this technique has shown its usefulness in the prenatal diagnosis of genetic diseases. Future applications may include other instances where small amounts of tissue are available (e.g. brain biopsies), or where only rare cells in a tissue sample contain the sequence of interest. (Editor's note: this method has been automated and even higher amplifications have been obtained. The reader is advised to consult with reference laboratories before attempting this procedure.)

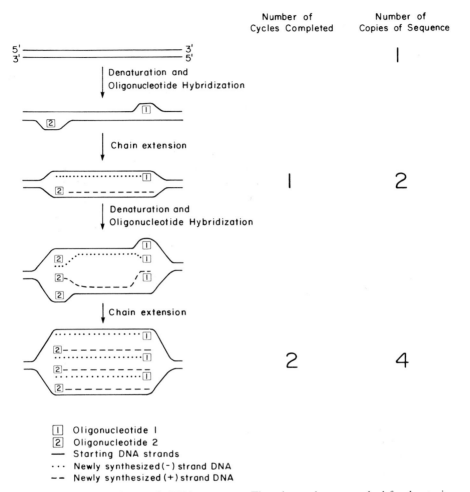

Fig. 4.5 Amplification of genomic DNA sequences. The scheme shows a method for denaturing genomic DNA, hybridizing oligonucleotides complementary to the 3' portion and identical to the 5' portion of a region to be amplified.[27] Chains are extended with the Klenow fragment of *E. coli* DNA polymerase I. The number of copies of the sequence of interest as a function of the number of cycles of denaturation, hybridization, and extension are shown assuming 100% efficiency.

Oligonucleotides can also be used for 'creating' genes which differ from a cloned prototype by as little as one specific nucleotide. To achieve this, the oligonucleotide is synthesized with a known mismatch to a specific region of a gene which exists as an insert in plasmid or bacteriophage. By initiating the synthesis of the gene and its vector on the purposely mismatched primer, and allowing the result of this reaction to replicate in the bacterial host, a clone with a site-specific mutation will be obtained.[44] See Chapter 11 Part One for details of one especially efficient approach to *in vitro* mutagenesis.

A number of approaches have been described for optimizing the efficiency of site-specific mutagenesis.[45–48] Random mutagenesis of a specific region can be achieved by using a single oligonucleotide preparation in which purposeful degeneracy is introduced at desired loci by using a mixture of the four nucleoside phosphoramites rather than a single one at the position in question.[49] Alternatively, a specific portion of a gene may be replaced with synthetic double stranded DNA, introducing a pre-designed alteration.[50]

Using these powerful techniques, it is possible to change, insert, or delete one or a few nucleotides within a cloned region of DNA. Site-specific mutagenesis can be applied to the study of RNA splicing by modifying one or a few nucleotides at splice donor, acceptor, or branch sites. The functional regions of an enzyme can be examined by truncating the protein through the introduction of termination codons or frame shifts into its coding region. Alternatively, single amino acid changes can be introduced into the signal sequence or active site. The effects of these alterations are amenable to testing in transfection assays which allow measurement of the function of the gene product.

SYNTHESIS AND USE OF RNA HYBRIDIZATION PROBES

Single stranded RNA transcripts of cloned DNA fragments have been extensively utilized for studies on RNA processing and as substrates for *in vitro* translation.[51–56] More recently, radiolabeled RNA copies of cloned genes have been applied as hybridization probes in a number of different assays.[57–59] The use of single stranded RNA molecules as hybridization probes has several clear theoretical advantages over the use of conventional nick translated DNA probes.[60,61] These advantages include:

1. the ability to use higher stringency washes because of the increased thermal stability of RNA : DNA hybrids versus the homologous DNA : DNA hybrids,
2. the use of DNA probes involves the establishment of a competition for binding between the sequence of interest and the two strands of the probe — no such competition is present when using single stranded RNA probes,
3. RNA probes can be labeled to very high specific activities.

These theoretical advantages result in a marked increase in sensitivity over that obtainable with conventional DNA probes. Unfortunately, these advantages have until recently remained mainly theoretical because of our inability to produce sufficient quantities of radioactively labeled RNA transcripts. However, with the recent discoveries of the bacteriophage SP6 and T_7 RNA polymerases, and the characterization of their strict promoter specificity, we now have in hand the reagents for the easy and efficient *in vitro* production of transcripts of recombinant genes.[62,63]

SP6 RNA polymerase is a DNA-dependent RNA polymerase obtained from *Salmonella typhimurium* infected with the bacteriophage SP6. It is a stable enzyme consisting of a single polypeptide chain of molecular weight 96 000 daltons.[62] This RNA polymerase initiates transcription exclusively at SP6 promoters. Utilizing this promoter-specificity, Melton et al constructed convenient cloning

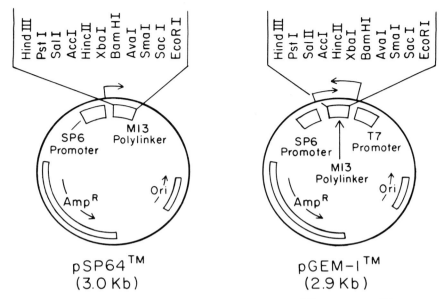

Fig. 4.6 Partial restriction map for the SP6 cloning vectors pSP64™ and pGEM-1™. Ori, origin of replication. Amp, ampicillin resistance gene. The arrows above the SP6 and T7 promoters indicate the start site and direction of transcription.

vectors which contained the SP6 promoter upstream from a polylinker cloning site (Fig. 4.6).[64] Use of these SP6 vectors allows the easy insertion of cloned DNA fragments downstream of the SP6 promoter. Linearization of these plasmids 3' to the cloned DNA fragment results in the production of a defined template for the subsequent efficient transcription by SP6 polymerase (Fig. 4.7). Repeated rounds of transcription by SP6 polymerase allow the production of up to microgram amounts of uniform single stranded RNA molecules. Incorporation of radioactive nucleotide triphosphates into the *in vitro* transcription reaction results in the production of a radiolabeled transcript which can have a specific activity of up to 1×10^9 cpm/μg. The ability to transcribe either strand of DNA by manipulating the orientation of the cloned fragment to the SP6 promoter can result in the production of either sense or antisense RNA.

Other RNA polymerases, most notably T_7 DNA dependent RNA polymerase isolated from bacteriophage T7 infected *Escherichia coli*, have similarly been used in the *in vitro* production of RNA probes.[65] Cloning vectors such as pGEM-1™ and pGEM-2™ have combined the use of both the SP6 and T7 promoters by flanking a polylinker sequence with each promoter (Fig. 4.6). This construct allows separate transcription of the coding and non-coding strand of the cloned DNA insert from the same plasmid by simply linearizing either 5' or 3' of the insert and utilizing the 3' or 5' promoter, respectively.

The introduction of SP6 snd T_7 cloning vectors has resulted in the rapid realization of the known theoretical advantages associated with the use of single

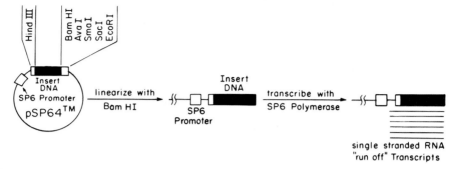

Fig. 4.7 Transcription from linearized pSP64™ cloning vector containing a hypothetical insert cloned into the Pst/Xbal sites of the M13 polylinker.

stranded RNA molecules as hybridization probes. The ease of preparation of these RNA probes, the high specific activity achievable, and their increased sensitivity in blot hybridization has led to the rapid assimilation of this technique into the repertoire of probe labeling options.

In this review we will detail an easy and efficient procedure for isolating transcription quality plasmids and the conditions required for the preparation and use of high specific activity radiolabeled single stranded RNA probes. Potential problems will be discussed as well as possible solutions.

Preparation of transcription quality plasmids

This protocol is a modification of the plasmid preparation procedure first introduced by Birnboim & Doly.[66] A single 3 ml culture should result in a sufficient yield of plasmid for up to 30 transcription reactions.

1. Inoculate 3 ml of L-Broth containing 35 μ/l ampicillin with a single bacterial colony. Incubate at 37°C overnight with vigorous shaking.
2. Centrifuge 1.5 ml of culture in a microfuge tube for 2 minutes.
3. Remove supernatant by aspiration and repeat Steps 2 and 3.
4. Resuspend the pellet by vortexing in 100 μl of freshly prepared 50 mM glucose, 25 mM Tris HCl, pH 8.0, 10 mM EDTA, 5 mg/ml lysozyme and incubate uncapped for 5 minutes at room temperature.
5. Add 200 μl of freshly prepared ice-cold 0.2 N NaOH, 1% SDS, mix by inverting several times and incubate on ice for 5 minutes.
6. Add 100 μl of ice-cold 3 M potassium acetate pH 5.2, mix gently by vortexing and incubate 5 minutes on ice.
7. Centrifuge for 5 minutes in a microfuge.
8. Transfer the supernatant to a clean microfuge tube, avoiding the white precipitate, and add 400 μl of a phenol/chloroform mix and centrifuge for 2 minutes.
9. Transfer upper aqueous phase to a clean microfuge tube and add 400 μl of chloroform, mix and centrifuge.
10. Transfer upper aqueous phase to a clean microfuge tube and add 1 ml of absolute ethanol and incubate at −20°C for 10 minutes.

11. Centrifuge for 5 minutes in a microfuge.
12. Remove the supernatant by careful aspiration and rinse the pellet with $-20°C$ 70% ethanol and dry the pellet.
13. Dissolve the pellet in 50 µl of 1 mM Tris HCl, pH 8.0, 0.1 mM EDTA. (A yield of approximately 10–15 µg of DNA should be expected.)

Linearization of plasmid
For end labeling or T_4 DNA polymerase reactions, the cloning vector containing the inserted template should be linearized with a restriction endonuclease which cuts only once within the plasmid and produces a 5' overhang or a blunt end. Restriction endonucleases such as Pst I and Sac I which produce 3' protruding ends should be avoided since use of these enzymes can result in transcripts which are copies of the entire plasmid DNA.[67] Conversion of 3' protruding ends to a blunt end through the use of T_4 DNA polymerase or the Klenow fragment of DNA polymerase should be performed if these enzymes cannot be avoided.

The restriction endonuclease digestion should be set up with the addition of RNase A (1 µg/ml) to remove the large amount of bacterial RNA contaminating the miniplasmid prep. This low level of RNase A is efficiently removed by subsequent treatments and therefore does not interfere with the later transcription by SP6 polymerase.

Following digestion, the reaction mixture is heated to 65°C for 10 minutes and then extracted with an equal volume of phenol 2×, phenol/chloroform 2× and chloroform 2×. The aqueous phase is then transferred to a clean microfuge tube and is precipitated by the addition of sodium acetate, pH 5.2, to 0.3 M and 2.5 volumes of absolute ethanol. The precipitate is washed with pre-chilled 70% ethanol, dried, and then resuspended in 250 µl of 10 mM Tris HCl, pH 8.0, 1 mM EDTA pH 8.0. The sample is then reprecipitated by the same approach and finally the dried pellet is resuspended in 50 µl of sterile water. Between 1 and 1.5 µl of this plasmid preparation (0.5–1.0 µg of DNA) can be used for the *in vitro* transcription reaction.

SP6 reaction conditions for high specific activity transcripts
Reactions are performed in a 20 µl volume in 40 mM Tris HCl, pH 7.5, 6 mM $MgCl_2$, 2 mM spermidine, 10 mM NaCl, 10 mM DDT, 1 unit/µl of the ribonuclease inhibitor RNasin®, 500 µM each of ATP, CTP, GTP, 12 µM α-[^{32}P]UTP (800 Ci/mM, Amersham), 0.5–1 µg of linearized DNA template and 10 units of SP6 RNA polymerase.* The reaction is incubated at 40°C for $1-1\frac{1}{2}$ hours.

This reaction condition typically results in the incorporation of greater than 80% of the free α-[^{32}P]UTP into transcripts. This results in the synthesis of approximately 250 ng of [^{32}P]RNA with a specific activity of greater than 2×10^8 cpm/µg. Free nucleotides can be conveniently separated from the incorporated nucleotides by centrifugation through a Sephadex G-100 column.[68]

---

* The reaction components are available in convenient kits from several commercial suppliers.

Use of this low concentration of UTP results in the frequent premature termination of transcription from templates greater than 2 kb. Full length transcripts are produced if smaller templates are used or if shorter incubation periods are possible. If large templates are required it is recommended that these be subcloned as separate fragments of 1–2 kb in size, and a mixture of these linearized templates used for the transcription reaction. Alternatively, a higher concentration of unlabeled UTP can be included, however, this will reduce the specific activity of the labeled transcript.

Hybridization protocol
A variety of different hybridization and washing conditions can be used and will depend primarily on the nature of the probe and the hybridization substrate. Typically, for the Southern blot analysis of single copy eukaryotic sequences, the filters are pre-hybridized and hybridized at 42°C in 50% formamide, 5 × SSC, 5 × Denhardt's solution (1 × = 0.02% BSA, 0.2% polyvinylpyrrolidone and 0.02% Ficoll), 0.2 M $NaPO_4$ pH 6.7, 10% dextran sulfate and 100 $\mu g/ml$ denatured salmon sperm DNA.

Following hybridization the filters are washed in 0.1 × SSC, 0.1% SDS for 15 minutes at room temperature. The filters are then washed twice in 0.1 × SSC, 0.1% SDS for 30 minutes at 58°C.

End labeling of DNA molecules
It is often advantageous to attach a radioactive label at either the 5′ or 3′ end of a DNA molecule that one wishes to use as a hybridization probe. By labeling only one end of the probe, one has the opportunity to derive information about the *position* of DNA sequences relative to the probe, as well as the presence and amount of the sequence. For example, one can end label a DNA probe, hybridize it to mRNA, and digest the mixture with S_1 nuclease. Since S_1 nuclease digests only single stranded regions of DNA, the only labeled, undigested DNA fragment obtained will be that part that is continuously hybridized to the mRNA from the end label at one end of the probe through to the first point at which significant mismatches between probe and the mRNA occur. Thus, one can use end labeling to map intron sequences in mRNA or DNA clones, areas of significant non-homology, the 5′ and 3′ ends of an mRNA molecule or a gene clone, etc.

The protocols described below are commonly used to label either the 5′ or 3′ end of a DNA molecule. The most efficient 5′ end labeling method involves the addition of a radioactive phosphate to the 5′ end of the strand by the enzyme polynucleotide kinase. Most naturally occurring DNA molecules already have a phosphate (non-radioactive, of course) at the 5′ end. Therefore, one must first remove this phosphate with calf intestinal phosphatase or bacterial alkaline phosphatase. These reagents, and protocols for their use, are readily available from a number of commercial suppliers. The most important step to remember when 'phosphatasing' DNA molecules is that the phosphatase enzyme must be removed or inactivated after the reaction. Otherwise, the residual phosphatase activity will very efficiently remove radioactive phosphates added during the kinase step. Calf intestinal phosphatase is far more sensitive to heat denaturization than bacterial alkaline phosphatase. Therefore, thorough heating of the sample at 65°C for 10–15 minutes will effectively inactivate the former, but not the latter, enzyme. Many laboratories follow heat inactivation with 2–3 phenol extractions as 'insurance'. Phenol extraction is absolutely essential if one uses bacterial alkaline phosphatase.

Synthetic oligonucleotides are usually labeled by the polynucleotide kinase reaction. Since synthetic oligonucleotides are invariably prepared with a terminal 5' hydroxyl group, prior phosphatase treatment is not necessary.

The principle of 3' end labeling is somewhat more complicated. No convenient enzyme that simply adds a phosphate group to the 3' end of DNA molecules is available. Rather, one prepares a restriction endonuclease fragment with a 5' overhanging end, such as Eco RI or Hind D III (cf Ch. 2). One then uses a DNA polymerase with a 3' to 5' polymerization direction, such as Klenow Fragment or T_4 polymerase, to 'fill in' the single stranded overhanging region with nucleotides, at least one of which is radioactive. This yields a blunt ended molecule with a radioactive label at, or extremely close to, the 3' end of the strand opposite the one containing the 5' overhang.

Alternatively, one can expose a blunt ended DNA molecule to brief digestion with T_4 polymerase in the absence of added nucleotide triphosphates. Under these conditions T_4 polymerase exhibits an exonuclease activity that works in the $3' \rightarrow 5'$ direction. If nucleotide triphosphates are then added back, the polymerase activity of T_4 DNA polymerase predominates, and the single stranded region created by the brief nuclease reaction will be filled in by the polymerase activity of the enzyme.

For restriction enzymes that leave a 3' overhanging end, the use of terminal transferase and a dideoxy nucleotide as label are now preferred to previous methods, which were somewhat cumbersome. This method is described by Dr Poncz in Chapter 9. In essence, terminal transferase will add nucleotides randomly to the 3' end of DNA chains at their 3' ends. One includes a dideoxy nucleotide, either labeled or as the appropriate next base beyond the labeled nucleotide. Chain extension is then limited to one or a few bases so that a true 3' label is obtained.

Protocol for 3' end labeling with Klenow Fragment
1. Digest up to 1 µg of DNA with the desired restriction enzyme in 25 µl of the appropriate buffer.
2. Add 2.0 µCi of the appropriate α-[^{32}P]dNTP.
3. Add 1.0 unit of the Klenow fragment of *E. coli* DNA polymerase, and incubate for 10 minutes at room temperature.
4. Either load the end labeled DNA directly onto a gel or separate the labeled DNA from unincorporated dNTPs by chromatography on or centrifugation through small columns of Sephadex G-50.

3' end labeling with 10 × T_4 polymerase buffer
0.33 M Tris acetate, pH 7.9
0.66 M potassium acetate
0.10 M magnesium acetate
5.0 mM dithiothreitol
1 mg/ml bovine serum albumin (BSA Pentax Fraction V)

The 10 × stock should be divided into small aliquots and stored frozen at $-20°C$.

1. Mix:
DNA	0.5–2.0 μg
10 × T_4 polymerase buffer	2 μl
H_2O	to 19 μl
2. Add the desired restriction enzyme and incubate for the appropriate time.
3. Add directly to the reaction 1 μl of a 2 mM solution of three of the four dNTPs (e.g. 2 mM dGTP, 2 mM dATP, 2 mM TTP).
4. Add approximately 2 μCi of an aqueous solution of the fourth dNTP, labeled with α-^{32}P (in the example given above, this would be α-[^{32}P]dCTP).
5. Add 1 μl (2.5 units) of T_4 DNA polymerase.
6. Incubate for 5 minutes at 37°C.
7. Add 1 μl of a 1 mM solution of the unlabeled, fourth dNTP (in this case, dCTP). Continue incubation for about 10 minutes.

In the presence of dNTPs, the 3' exonuclease activity of the T_4 DNA polymerase on double stranded DNA is masked by the polymerization reaction. Thus, the product of the protocol described above will be a blunt ended molecule with labeled nucleotides incorporated at or very near the termini of the DNA.

Protocol for 5' end labeling
1. Mix:
phosphorylated DNA, 5' ends or DNA oligonucleotide	1–50 pmol
10 × kinase buffer	5 λ
γ-[^{32}P]ATP (specific activity = 3000 Ci/mmol)	150 μg
T_4 polynucleotide kinase	10–20 units
H_2O	to 50 λ

   Incubate at 37°C for 60 minutes.
   10 × kinase buffer: 0.5 M Tris Cl, pH 7.6
   0.1 M $MgCl_2$
   1.0 mM spermidine
   50.0 mM dithiothreitol
2. Add 2 λ of 0.5 M EDTA pH 8.0. Phenol/chloroform extract. Add $\frac{1}{10}$ volume 3 M sodium acetate pH 5.5 and 2 volumes of ethanol to precipitate. Store at $-70°C$ for 1 hour or $-20°C$ overnight.
3. Spin out DNA, dry, redissolve in 100 λ TE.
4. Purify DNA from unincorporated γ[^{32}P]ATP by passage over a G-50 medium spun column.
5. Quantitate amount of radioactive probe by TCA precipitation.

CONCLUSION

Use of the above procedures results in the easy preparation of transcription quality plasmids and the efficient production of high specific activity probes. This labeling system eliminates the need for large scale plasmid preparations involving cesium chloride density gradient centrifugation and the necessity to isolate inserts away from the cloning vector. This labeling system has been rapidly incorporated into the molecular biologists' armamentarium and has found frequent usage as a means of labeling probes for Southern and Northern hybridization analysis. In addition, the high specific activity of these probes and their increased hybridization stability over double stranded DNA probes has made them uniquely suited for use in *in situ* hybridization of tissue sections.[69]

REFERENCES

1 Singer-Sam J, Simmer R L, Keith D H et al. Isolation of a cDNA clone for human X-linked 3-phosphoglycerate kinase by use of a mixture of synthetic oligodeoxyribonucleotides as a detection probe. Proc Natl Acad Sci USA 1983; 80: 802–806.
2 Takizawa T, Huang I-Y, Ikuta T, Yoshida A. Human glucose-6-phosphate dehydrogenase: Primary structure and cDNA cloning. Proc Natl Acad Sci USA 1986; 83: 4157–4161.
3 Langer P, Waldrop A A, Ward D C. Enzymatic synthesis of biotin-labeled polynucleotides: Novel nucleic acid affinity probes. Proc Natl Acad Sci USA 1981; 78: 6633–6637.
4 Brigati D J, Meyerson D, Leary J J et al. Detection of viral genomes in cultured cells and paraffin-embedded tissue sections using biotin-labeled hybridization probes. Virology 1983; 126: 32–50.
5 Leary J J, Brigati D J, Ward D C. Rapid and sensitive colorimetric method for visualizing biotin-labeled DNA probes hybridized to DNA or RNA immobilized on nitrocellulose: Bio-blots. Proc Natl Acad Sci USA 1983; 80: 4045–4049.
6 Rigby P W J, Dieckmann M, Rhodes C, Berg P. Labeling deoxyribonucleic acid to high specific activity *in vitro* by nick translation with DNA polymerase I. J Mol Biol 1977; 113: 237–251.
7 Balmain A, Birnie G D. Nick translation of mammalian DNA. Biochim Biophys Acta 1979; 561: 155–166.
8 O'Farrell P H, Kutter E, Nakanishi M. A restriction map of the bacteriophage T4 genome. Mgg 1980; 179: 421–435.
9 Hallberg M D, Englund P T. Specific labeling of 3' termini with T4 DNA polymerase. Methods Enzymol 1980; 65: 39–42.
10 Feinberg A P, Vogelstein B. A technique for radiolabeling DNA restriction endonuclease fragments to high specific activity. Anal Biochem 1983; 132: 6–13.
11 Feinberg A P, Vogelstein B. Addendum to above article. Anal Biochem 1984; 137: 266–267.
12 Narang S A, Brousseau R, Hsiung H M, Michniewicz J J. Chemical synthesis of deoxyoligonucleotides by the modified triester method. Methods Enzymol 1980; 65: 610–620.
13 Reese C B. The chemical synthesis of oligo- and polynucleotides by the phosphotriester approach. Tetrahedron 1980; 34: 3143–3179.
14 Ohtsuka E, Ilehara M, Soll D. Recent developments in the chemical synthesis of polynucleotides. Nucleic Acids Res 1982; 10: 6553–6570.
15 Narang S A. DNA synthesis. Tetrahedron 1983; 39: 3–22.
16 Itakura K, Rossi J J, Wallace R B. Synthesis and use of synthetic oligonucleotides. Annu Rev Biochem 1984; 53: 323–356.
17 Tu C-PD, Wu R. Sequence analysis of short DNA fragments. Methods Enzymol 1980; 65: 620–638.
18 Chaconas G, van de Sande J. 5'-^{32}P-labeling of RNA and DNA restriction fragments. Methods Enzymol 1980; 65: 75–85.

19 Yousaf S I, Carroll A R, Clarke B E. A new and improved method for 3'-end labeling DNA using α-^{32}P-ddATP. Gene 1984; 27: 309–313.
20 Tu C-PD, Cohen S N. 3'-end labeling of DNA with [α-^{32}P]cordycepin-5'-triphosphate. Gene 1980; 10: 177–183.
21. Zullo J N, Cochran B H, Huang A S, Stiles C D. Platelet-derived growth factor and double-stranded ribonucleic acids stimulate expression of the same genes in 3T3 cells. Cell 1985; 43: 793–800.
22 Roychoudhoury R, Wu R. Terminal transferase-catalyzed addition of nucleotides to the 3' termini of DNA. Methods Enzymol 1980; 65: 43–62.
23 Maxam A, Gilbert W. Sequencing end-labeled DNA with base-specific chemical cleavages. Methods Enzymol 1980; 65: 499–560.
24 Collins M L, Hunsaker W R. Improved hybridization assays employing tailed oligonucleotide probes: A direct comparison with 5' end labeled oligonucleotide probes and nick translated plasmid probes. Anal Biochem 1985; 151: 211–224.
25 Ratner L, Josephs S F, Starcich B, Shaw G M, Hahn B, Gallo R C, Wong-Staal F. Nucleotide sequence analysis of a variant human T-cell leukemia virus (HTLV-Ib) provirus with a deletion in pX-I. J Virol 1985; 54: 781–790.
26 Conner B J, Reyes A A, Morin C, Itakura K, Teplitz R L, Wallace R B. Detection of sickle cell β-globin allele by hybridization with synthetic oligonucleotides. Proc Natl Acad Sci USA 1983; 80: 278–282.
27 Piratsu M, Kan Y W, Cao A, Conner B J, Teplitz R L, Wallace R B. Prenatal diagnosis of β-thalassemia: Detection of a single nucleotide mutation in DNA. N Engl J Med 1983; 309: 284–287.
28 Orkin S H, Markham A F, Kazazian H H. Direct demonstration of the common Mediterranean β-thalassemia gene with synthetic DNA probe. J Clin Invest 1983; 71: 775–779.
29 Kidd V J, Wallace R B, Itakura K, Woo S L C. α1-antitrypsin deficiency detection by direct analysis of the mutation in the gene. Nature 1983; 304: 230–234.
30 Grantham R, Gautier C, Gouy M, Jacobzone M, Mercier R. Codon catalog usage is a genome strategy modulated for gene expressivity. Nucleic Acids Res 1981; 9: 443–473.
31 Yang J, Ye J, Wallace D C. Computer selection of oligonucleotide probes from amino acid sequences for use in gene library screening. Nucleic Acids Res 1984; 12: 837–843.
32 Wood W I, Gitschier J, Lasky L A, Lawn R M. Base composition-independent hybridization in tetramethylammonium chloride: A method for oligonucleotide screening of highly complex gene libraries. Proc Natl Acad Sci USA 1985; 82: 1585–1588.
33 Sanger F, Nicklen S, Coulson A R. DNA sequencing with chain terminating inhibitors. Proc Natl Acad Sci USA 1977; 74: 5463–5467.
34 Hong G F. Sequencing of large double-stranded DNA using the dideoxy sequence technique. Biosci Rep 1982; 2: 907–912.
35 Dente L, Cesareni G, Cortese R. pEMBL: A new family of single-stranded plasmids. Nucleic Acids Res 1983; 11: 1645.
36 Chan E Y, Seeburg P H. Supercoil sequencing: A fast and simple method for sequencing plasmid DNA. DNA 1985; 4: 165–170.
37 Griffiths C, Berek C, Kaartinen M, Milstein C. Somatic mutation and the maturation of immune response to 2-phenyl oxazolone. Nature 1984; 312: 271–275.
38 Prchal J T, Cashman D P, Kan Y W. Hemoglobin Long Island is caused by a single mutation resulting in a failure to cleave amino-terminal methionine. Proc Natl Acad Sci USA 1986; 83: 24–27.
39 Ghosh P, Reddy V B, Piatak M, Lebowitz P, Weissman S M. Determination of RNA sequences by primer directed synthesis and sequencing of their cDNA transcripts. Methods Enzymol 1980; 65: 580–595.
40 Winter G, Fields S, Gait M J. The use of synthetic oligodeoxynucleotide primers in cloning and sequencing segment 8 of influenza virus (A/PR/8/34). Nucleic Acids Res 1981; 9: 237–245.
41 Goeddel D V, Shepard H M, Yelverton E, Leung D, Crea R. Synthesis of human fibroblast interferon by E. coli. Nucleic Acids Res 1980; 8: 4057–4074.
42 Walter P, Green S, Greene G et al. Cloning of human estrogen receptor cDNA. Proc Natl Acad Sci USA 1985; 82: 7889–7893.
43 Saiki R K, Scharf S, Faloona F, Mullis K B, Horn G T, Erlich H A, Arnheim N. Enzymatic amplification of β-globin genomic sequences and restriction site analysis for diagnosis of sickle cell anemia. Science 1985; 230: 1350–1354.

44 Zoller M J, Smith M. Oligonucleotide-directed mutagenesis of DNA fragments cloned into M13 vectors. Methods Enzymol 1983; 100: 468–500.
45 Gilliam S, Astell C R, Smith M. Specific mutagenesis using oligodeoxyribonucleotides: Isolation of a phenotypically silent IX174 mutant, with a specific nucleotide deletion at very high efficiency. Gene 1980; 12: 129–137.
46 Norris K, Norris F, Christianson L, Fiil N. Efficient site-directed mutagenesis by simultaneous use of two primers. Nucleic Acids Res 1983; 11: 5103–5112.
47 Oostra B A, Harvey R, Ely B K, Markham A F, Smith A E. Transforming activity of polyoma virus middle-T antigen probed by site-directed mutagenesis. Nature 1983; 304: 456–459.
48 Kramer W, Schughart K, Fritz H-J. Directed mutagenesis of DNA cloned in filamentous phage: Influence of hemimethylated GATC sites on marker recovery from restriction fragments. Nucleic Acids Res 1982; 10: 6475–6485.
49 Hutchison C A, Nordeen S K, Vogt K, Edgell M H. A complete library of point substitution mutations in the glucocorticoid response element of mouse mammary tumor virus. Proc Natl Acad Sci USA 1986; 83: 710–714.
50 Ferretti L, Karnik S S, Khorana G, Nassal M, Oprian D D. Total synthesis of a gene for bovine rhodopsin. Proc Natl Acad Sci USA 1986; 83: 599–603.
51 Kruger K, Grabowski P J, Zaug A J, Sands J, Gottschling D E, Cech T R. Self-splicing RNA: Autoexcision and autocyclization of the ribosomal RNA intervening sequence of tetrahymena. Cell 1982; 31: 147–157.
52 Padgett R A, Hardy S F, Sharp P A. Splicing of adenovirus RNA in a cell-free transcription system. Proc Natl Acad Sci USA 1983; 80: 5230–5234.
53 Green M R, Maniatis T, Melton D A. Human γ-globin pre-mRNA synthesized in vitro is accurately spliced in xenopus oocyte nuclei. Cell 1983; 32: 681–694.
54 Hernandez N, Keller W. Splicing of in vitro synthesized messenger RNA precursors in HeLa cell extracts. Cell 1983; 35: 89–99.
55 Krieg P A, Melton D A. Formation of the 3' end of histone mRNA by post-transcriptional processing. Nature 1984; 308: 203–206.
56 Krainer A R, Maniatis T, Ruskin B, Green M R. Normal and mutant human β-globin pre-mRNAs are faithfully and efficiently spliced in vitro. Cell 1984; 36: 993–1005.
57 Brahic M, Haase A T. Detection of viral sequences of low reiteration frequency by in situ hybridization. Proc Natl Acad Sci USA 1978; 6125–6129.
58 Diaz M, Barsacchi-Pelane G, Mahon K, Gall J. Transcripts from both strands of a satellite DNA occur on Lampbrush chromosome loops of the newt notophthalmus. Cell 1981; 24: 649–659.
59 Lynn D A, Angerer L M, Bruskin A M, Klein W H, Angerer R C. Localization of a family of mRNAs in a single cell type and its precursors in sea urchin embryos. Proc Natl Acad Sci USA 1983; 80: 2656–2660.
60 Zinn K, DeMaoi D, Maniatis T. Identification of two distinct regulatory regions adjacent to the human β-interferon gene. Cell 1983; 34: 865–879.
61 Church G M, Gilbert W. Genomic sequencing. Proc Natl Acad Sci USA 1984; 81: 1991–1995.
62 Butler E T, Chamberlin M J. Bacteriophage SP6-specific RNA polymerase. I. Isolation and characterization of the enzyme. J Biol Chem 1982; 257: 5772–5778.
63 Kassavetis G A, Butler E T, Roulland D, Chamberlin M J. Bacteriophage SP6-specific RNA polymerase. II. Mapping of SP6 DNA and selective in vitro transcription. J Biol Chem 1982; 257: 5779–5788.
64. Melton D A, Krieg P A, Rebagliati M R, Maniatis T, Zinn K, Green M R. Efficient in vitro synthesis of biologically active RNA and RNA hybridization probes from plasmids containing a bacteriophage SP6 promoter. Nucleic Acids Res 1984; 12: 7035–7056.
65 Davanloo P, Rosenberg A H, Dunn J J, Studier F W. Cloning and expression of the gene for bacteriophage T7 RNA polymerase. Proc Natl Acad Sci USA 1984; 81: 2035–2039.
66 Birnboim H C, Doly J. A rapid alkaline extraction procedure for screening recombinant plasmid DNA. Nucleic Acids Res 1979; 7: 1513–1523.
67 Schenborn E T, Mierendorf R C Jr. A novel transcription property of SP6 and T7 RNA polymerases: Dependence on template structure. Nucleic Acids Res 1985; 13: 6223–6236.
68 Maniatis T, Frilsch E F, Sambrook J. Molecular cloning: A laboratory manual. Cold Spring Harbor, NY: Cold Spring Harbor Laboratory, Cold Spring Harbor Press, 1982: p 464–465.
69 Cox K H, Delson D V, Angerer L M, Angerer R C. Detection of mRNAs in sea urchin embryos by in situ hybridization using asymmetric RNA probes. Dev Biol 1984; 101: 485–502.

5
cDNA and genomic cloning
E. V. Prochownik   F. S. Collins

INTRODUCTION

Both the theoretical and practical aspects of modern molecular biology are based on the ability to propagate defined eucaryotic DNA sequences in procaryotic hosts (commonly referred to as cloning). These DNA sequences may be either DNA copies of mRNAs or actual genomic segments. They are inserted into plasmid or bacteriophage vectors such that they are replicated as non-deleterious, neutral DNA fragments.

Methods for the preparation of eucaryotic cDNA and genomic sequences and their insertion into procaryotic cloning vectors form the backbone of contemporary molecular biology. In this chapter we present some of the methods which have served our laboratories in a reproducible manner. For excellent discussions of cloning and the various vectors employed, the reader is referred to Maniatis et al.[1]

LARGE-SCALE GROWTH OF BACTERIA FOR PLASMID ISOLATION

The preparation of bacterial plasmids is an essential technique of molecular biology. Most published procedures for the growth of plasmids include an amplification step with chloramphenicol to increase the copy number. While this results in quite high yields of plasmid DNA, it suffers from the drawback that the culture must be incubated for several hours and monitored closely before the chloramphenicol is added. In the procedure described below, the amplification step has been eliminated without significantly affecting the yield of plasmid DNA. Thus, we find it quite feasible to inoculate plasmid minipreps in the morning, prepare and analyze the DNA by late afternoon and then inoculate the appropriate large-scale culture for overnight growth with the remainder of the desired miniprep culture.

Required solutions:
  10 × M9 salts (58 g/l Na_2HPO_4, 30 g/l KH_2PO_4, 5 g/l NaCl and 10 g/l NH_4Cl)
  0.25 g/ml uridine filter sterilized
  1 M $MgCl_2$
  stock solution of appropriate antibiotic (25 mg/ml of ampicillin, 10 mg/ml tetracycline)

Procedure: Prepare 1 litre of 1 × M9 salts in H_2O. Add 5 g casamino acids. Autoclave. Cool to room temperature. The sterile solution can be stored at room temperature for several weeks. When ready to inoculate bacteria, add 4 ml of uridine, 1 ml of $MgCl_2$ and 1 ml of antibiotic. Add bacteria and shake vigorously overnight at 37°C in a well-aerated flask. The following day, prepare the plasmid by standard methods (see Chs. 4, 6, 9, 13, or reference[1] for alternative methods).

GROWTH OF BACTERIOPHAGE FOR DNA ISOLATION

Unfortunately, bacteriophage DNA is less straightforward to prepare than is plasmid, which is apparent from the large number of protocols in the literature which attempt to solve various problems with phage DNA isolation. The following protocols have given good results in our hands, but even after many uses occasionally fail to give good yields of DNA. The reason usually relates to the ratio of phage to host bacteria at the time of inoculation. Too many phage results in premature lysis of the culture; too few results in no lysis at all. Therefore, it is often helpful to include a range of multiplicities of infection bracketing the expected optimum, and then one can choose the best-lysed culture (indicated by floating stringy particulate material) for DNA isolation.

Small scale minilysates
1. Pick desired plaque from a nutrient agar plate into 1 ml PSB (Phosphate buffered saline). If you have a titered stock, start with about 10^4 to 10^5 pfu.
2. After standing 1 hour, add 1 μl to 0.2 ml of fresh overnight culture of host. In a separate tube, add 10 μl to 0.2 ml of overnight culture.
3. Incubate at 37°C for 15 minutes, then add 10 ml NZYM (see glossary). Shake at 37°C overnight.
4. Pick the culture which shows the best lysis. Add a few drops of chloroform, shake, let stand 5 minutes, then spin at 300 rpm in Beckman TJ-6 for 10 minutes. Pour supernatant into a fresh tube.
5. Save 5 ml of this phage as a stock. To the remainder add 5 μg DNase and 5 μg RNase. Incubate at 37°C for one hour.
6. Add 200 μl proteinase K (25 mg/ml); incubate at 37°C for 15 minutes. Then add 250 μl 0.5 M EDTA and 250 μl 10% SDS. Incubate at 65°C for one hour. Spin out any floating particulates at 3000 rpm for 10 minutes.
7. Transfer to SW41 tubes. Add 1.0 ml 5 M LiCl, mix well, and then add 6 ml of cold isopropanol. Mix, spin at 23 000 rpm for 20 minutes.
8. Resuspend pellet in 400 μl TE and transfer to an eppendorf tube. Phenol extract twice, ether extract once (see Ch. 2), add 40 μl of 3 M sodium acetate and 800 μl of cold ethanol; chill at -20°C for 1 hour; spin, wash pellet with cold 70% ethanol, dry and resuspend in 50 μl TE.
9. Expected yield is about 5–20 μg. Use 1–4 μl in restriction digests, which should include RNase.

Large scale preps

Preparing large quantities of phage DNA can also be fraught with difficulties. The following protocol, which is a blend of many described in the literature, has worked well for most phage we have tried. Though it is long, it gives reliable results so long as visible lysis occurs in the original culture.

1. Prepare sterile quantity of NZYM (1 liter) for each phage prep to be made. Add 5×10^5 to 5×10^6 pfu of a titered stock of the phage to 0.3 ml of an overnight culture of the host (e.g. LE392★), which has been grown in NZYM with 0.4% maltose to induce phage receptors. The exact titer which gives best lysis will vary from phage to phage, and is crucial to obtaining a high yield. In general, phage which give small plaques will require a higher inoculum for optimum lysis. For a new phage it is best to try several different amounts of phage, and pick the culture which shows the best lysis. After incubating phage and host for 15 minutes at 37°C, add to the NZYM broth.
2. Incubate at 37°C with vigorous shaking (300 rpm) until lysis occurs (8–12 hours). For Charon phages this can require overnight shaking.
3. Add 10 ml chloroform to each liter to complete the lysis. Shake for 10 minutes at 37°C. Add 30 g/l NaCl, 200 μg DNase, and 1 mg RNase. The nucleases are very important to break up clumps of phage which are held together by bacterial nucleic acids. Shake another hour at 37°C. Spin out cell debris at 7500 rpm for 30 minutes at 4°C.
4. Remove chloroform from bottom of collected supernatant, add 4 ml 3 M $MgSO_4$. Add 120 g polyethylene glycol 6000 (PEG) per liter of supernatant to precipitate phage. Stir overnight in cold room.
5. Recover PEG pellet at 7500 rpm at 4°C for 20 minutes. The pellet will be glassy and spread out along the wall of the centrifuge bottle. Using rubber policemen, gently resuspend pellets in a total of 16 ml SM phage buffer (100 mM NaCl, 10 mM $MgSO_4$, 10 mM Tris HCl, pH 7.4). Be sure to wash walls of centrifuge bottles thoroughly.
6. Extract once with 15 ml chloroform to remove PEG. Shake well and spin at 3000 rpm in Beckman TJ-6 for 15 minutes. A dense interface will be present.
7. Avoiding interface, take off upper phase, and add 0.75 g/ml solid CsCl to phage suspension. Mix gently.
8. Load into SW41 ultracentrifuge ultraclear tubes. Spin at 35 000 rpm for 20 hours at 4°C. Allow the centrifuge to stop without braking.
9. Using a black background to visualize the phage, remove blue band of phage particles (which will be about two-thirds of the way down the tube), using 20 gauge needle. Avoid other bands of contaminated material when removing blue band.

★ Editor's note: LE 392 is an *Escherichia coli* strain readily available from bacteriology laboratories and commercial sources.

10. Remove cesium by dialysis against two changes of 50 mM Tris, pH 7.4, 10 mM NaCl, 10 mM $MgCl_2$ (one hour and two liters each).
11. Transfer to 15 ml conical tube. Break open phage by addition of 1/25 volume of 0.5 M EDTA, 1/500 volume 25 mg/ml proteinase K, and 1/20 volume 10% SDS. Incubate at 65°C for 60 minutes.
12. Extract twice with phenol/chloroform/isoamyl alcohol (see Ch. 2), and once with ether. Ethanol precipitate, and resuspend in 500 µl TE. Concentration should be approximately 1 µg/µl, which can be determined by OD measurements or gel electrophoresis against known standards.

CONSTRUCTION OF cDNA LIBRARIES

Two general methods are currently available for constructing cDNA libraries. The first, more traditional, approach involves the insertion of dCMP-tailed double stranded cDNAs into a plasmid vector which has been linearized at its unique Pst I site and tailed with dGMP residues. With the second method, molecular linkers are added to the ends of ds cDNAs followed by ligation into λ-bacteriophage-derived vectors and *in vitro* DNA packaging.[2] This second method affords two major advantages. First, the high efficiency of *in vitro* packaging often results in the generation of several million independent recombinants per µg of ds cDNA. Second, the phage vectors employed allow for the actual expression of properly inserted cDNA sequences, thus permitting the screening of recombinant clones by immunological methods and obviating the need for oligonucleotide probes based on protein sequences. This method is therefore particularly suited to searching for non-abundant transcripts for which protein sequence data is lacking. On the other hand, use of this method makes the characterization of cDNA clones of interest more burdensome due to the greater complexity of the λ vector versus plasmid vectors. Thus, cDNA clones often must be recloned in plasmid vectors to allow for easier restriction mapping.

We present below methods for constructing cDNA libraries using each of the above methods.

Method I: preparation of dCMP-tailed ds cDNA in plasmid vectors

First strand synthesis
The reaction below is based on the use of 5 µg of poly-A^+ mRNA as starting material. Based on our experience, the final yield of size-selected, dCMP-tailed ds cDNA will be 0.5–1 µg. Add the reagents below to an autoclaved and siliconized 1.5 ml eppendorf tube. All reagents should be prepared in DEPC-treated water and stored in sterile containers. All glassware should be baked.

10 mM dATP	2.5 μl
10 mM dGTP	2.5 μl
10 mM dTTP	2.5 μl
2 mM dCTP	2.5 μl
5 × RT Buffer (250 mM Tris HCl, pH 8.2; 250 mM KCl; 40 mM MgCl$_2$)	10 μl
200 mM dithiotreitol	2.5 μl
oligo-dT (600 μg/ml; Collaborative Research)	4 μl
[^{32}P]dCTP (sp. act. 3000 Ci/mmol)	10 μl (100 μCi)
RNasin ribonuclease inhibitor (Promega Biotech)	5 units
poly-A$^+$ mRNA	5 μg
AMV reverse transcriptase (Life Sciences)	10 units
H$_2$O	to a final volume of 50 μl

Incubate at 42° for 60 minutes. Terminate the reaction by adding 2 μ of 500 mM EDTA, pH 8.0. TCA precipitate* 1–2 μl to determine the efficiency of the reaction, which is usually 15–25%, rarely as high as 40%.

Extract the reaction twice with equilibrated phenol and twice with ether. Precipitate the reaction three times to remove unincorporated [^{32}P]dCTP. At this point, we have found it beneficial to introduce a size selection step. We originally utilized alkaline sucrose gradient centrifugation with excellent results.[3] However, the procedure is time-consuming, requiring a 24–48 hour spin. More recently, we have utilized chromatography over Sepharose CL-4B columns with nearly the same success. We do not recommend sizing cDNA through alkaline agarose gels since agarose contaminants can severely inhibit the Klenow DNA polymerase used in the next step.

After the last precipitation of the cDNA, re-dissolve the pellet in 100–200 μl of H$_2$O. Dialyze this overnight at 4°C against 200 ml of 30 mM NaOH, 2 mM EDTA. The next morning, dialyze for 2–3 hours against several changes of TE. Schleicher & Schuell collodion bags work well for this and do not bind DNA.

The Sepharose CL-4B column should be prepared in a siliconized Pasteur pipette plugged with glass wool and washed extensively with TE. The [^{32}P] cDNA is then added and the column washed with individual 50 μl aliquots of TE. Individual fractions are monitored with a Geiger counter. 2000–5000 counts from each fraction may then be run on an alkaline agarose gel to determine the size distribution of the cDNA. We generally pool fractions containing cDNAs > 500 nt in length (approx. 60% of the total).

---

* Editor's note: add the aliquot to 0.2 ml cold TE buffer (10 mM Tris HCl, pH 7.5, 1 mM EDTA, pH 7.0). Remove 5 μl and spot this directly on a dry filter. Add 20 μl 100% trichloroacetic acid (TCA) to remaining 195 μl. Stand on ice for 20 minutes. Collect precipitate (invisible) by vacuum filtration through a nitrocellulose filter. Wash the filter with 10% TCA. Dry under a heat lamp. Count in scintillation counter. The % of nucleotide incorporated

$$= \frac{\text{cpm on TCA precipitated Filter}}{40 \times \text{cpm on spotted filter}} \times 100$$

Second strand synthesis
We have found that Klenow DNA polymerase gives the best results at ssDNA concentrations of 2–5 µg/ml. For the following reaction, we assume 1 µg of starting ssDNA.

10 mM of each dNTP	12.5 µl each (total of 50 µl)
700 mM KCl	25 µl
5 mM β-mercaptoethanol	25 µl
10 × Klenow buffer (300 mM Tris HCl, pH 7.5, 40 mM MgCl^2)	25 µl
ss[^{32}P]cDNA	1 µg
Klenow DNA Polymerase (Boehringer-Mannheim)	50 units
H$_2$O	to a final volume of 250 µl

Incubate at 18°C for 16–20 hours.
Extract with phenol and ether as above. Dialyze exhaustively against water in a collodion bag, since free deoxynucleoside triphosphates may interfere with the subsequent S$_1$ nuclease step.

S$_1$ nuclease reaction
The volume of the Klenow reaction will increase somewhat during dialysis. Add 1/10 volume of 10 × S$_1$ Buffer (3 M NaCl; 300 mM Na acetate; pH 4,5; 100 mM ZnCl$_2$) and S$_1$ nuclease (Sigma) to a final concentration of 10 U/ml. Incubate at 37°C. Periodically remove 10–50 µl aliquots and TCA precipitate to monitor the extent of digestion. 60–80% of the input counts should be S$_1$-resistant. The reaction generally plateaus after about 10 minutes and we terminate it after 30 minutes by adding EDTA to a final concentration of 10 mM. It is important not to add too much S$_1$ nuclease or to incubate for too long since this will result in internal nicking of the ds cDNA and in fewer perfectly blunt ended molecules.

The S$_1$ reaction is immediately extracted with phenol and dialyzed against water at room temperature for 5–6 hours. The DNA is then concentrated by centrifugation (best performed in an ultracentrifuge where recovery of small amounts of DNA will be complete).

At this point, we again size select the ds cDNA, once again monitoring individual fractions and combining only those containing ds cDNAs > 500 bp in length.

Tailing ds cDNA with terminal transferase (TdT)
The reaction below works most efficiently with cDNA end concentrations of 20 nM. It should be remembered that lower molecular weight species of cDNA will contribute a disproportionately large number of ends and will, therefore, be tailed in preference to higher molecular weight species. It is precisely for this reason that we feel two size selection steps are critical to obtaining the longest clonable cDNA inserts.

The conditions below are for 1 µg of ds cDNA. Expect an average of 10–20 dCMP residues per cDNA end:

2 × cacodylate buffer (250 mM Na cacodylate, pH 7.19, 0.2 mM DTT)	200 µl
50 µM dCTP	40 µl
25 mg/ml nuclease-free bovine serum albumin (BRL (Bethesda Research Laboratories))	8 µl
ds cDNA (50 ng/µl)	20 µl
10 mM CoCl$_2$*	20 µl
TdT (BRL)	300 units
H$_2$O	To final vol of 400 µl

* Add just before enzyme or Co^{2+} will precipitate.

Incubate the above reaction at 20°C for 20 minutes. Phenol extract and precipitate. Lyophilize the pellet and then re-dissolve it in 50 µl of H$_2$O.

Annealing of GMP-tailed plasmid DNA and dCMP-tailed cDNA inserts

A TdT reaction similar to that above can be used to add dGMP residues to Pst I-digested pBR322. However, tailed plasmid DNAs of high quality can be readily obtained commercially (New England Nuclear, BRL).

Annealing reactions are most conveniently performed in 10 µl sealed glass capillary tubes. A cDNA to vector ratio of 1:1 is optimal in most cases. However, each preparation of cDNA should be test-annealed with varying amounts of plasmid in order to establish optimal conditions. Below is a sample reaction:

dCMP-tailed cDNA	1 µl (5 ng).
dGMP-tailed plasmid	1 µl (20 ng)
10 × annealing buffer (1 M NaCl; 100 mM Tris HCl, pH 7.5; 10 mM EDTA)	1 µl
H$_2$O	7 µl

Heat the reaction at 68°C for 8 minutes. Transfer to a 42°C water bath for 2 hours. Turn off the water bath and leave the annealing reaction overnight. Transform competent *E. coli* the following day.

Transformation of competent E. coli

It is imperative that bacterial cells of high-transforming efficiency be used for this step. Using the procedure of Hanahan, we have routinely been able to achieve transformation efficiencies in excess of 5×10^7 colonies/µg of supercoiled pBR322 DNA using *E. coli* strain MC1061. This translates to $> 10^5$ colonies/µg of ds-tailed cDNA. Recently, several companies have offered frozen competent cells with a reported transformation efficiency in excess of 10^8 colonies/µg of pBR322 DNA.

Plating and storage of recombinant colonies
Recombinant colonies should be grown on nitrocellulose filters which in turn are placed directly onto an antibiotic-containing agar plate. It is important that the plates be 'aged' by allowing them to dry out a bit. This is most conveniently done by placing a freshly poured set of uncovered plates in a laminar flow hood for 1–2 hours or until the agar surface takes on a slightly wrinkled appearance. This will prevent spreading of the colonies after plating.

The initial plating represents the unamplified library which should consist of at least 2×10^5 independent clones. Storage is most conveniently accomplished by scraping the filters into 25–50 ml of 25% glycerol in L-Broth. This concentrated bacterial suspension may then be stored indefinitely at $-80°C$ in small aliquots. For actual screening, the library is thawed, replated on nitrocellulose filters at 2000–4000 colonies per filter and then replica plated and probed.

Method II: preparation of ds cDNA in λgt 10 or λgt 11 vectors

First strand synthesis
The same considerations apply for this method as for Method I. In this method, we again recommend starting with 5 µg of poly-A$^+$ mRNA. Synthesis of the first strand is as described for Method I except that no [^{32}P]dCTP is added to the reaction, and the amount of unlabeled dCTP is raised to equal that of the other three dNTPs. The efficiency of the reaction can be monitored by removing a 2.5 µl aliquot and adding it to a separate tube containing 0.5 µl of [^{32}P]dCTP (3000 Ci/mmol). Phenol extract and precipitate the completed reaction. Wash the pellet in 70% ethanol and lyophilize it.

Second strand synthesis
Redissolve the cDNA:RNA pellet in 293 µl of H$_2$O. Then add the following:

5 × second strand buffer (100 mM Tris HCl, pH 7.5, 25 mM MgCl$_2$, 50 mM (NH$_4$)$_2$SO$_4$, 500 mM KCl, 50 mM dithiothreitol, 250 µg/ml BSA)	80 µl
10 mM dNTPs	6.4 µl (1.6 µl of each)
15 mM β-NAD (freshly prepared)	4 µl
RNase H (2 U/l)	2 µl
E. coli DNA Ligase (5 U/l)	1 µl
E. coli DNA polymerase I (9.8 U/l)	10 µl
[^{32}P]dCTP (~3000 Ci/mmol)	2 µl
	Final volume = 400 µl

Incubate at 14°C overnight.

TCA precipitate 2 µl to determine the efficiency of the second strand reaction. Phenol and ether extract the reaction followed by ethanol precipitation. Reprecipitate the pellet, wash with 70% ethanol and lyophilize.

Blunting reaction

In order to efficiently add the Eco RI linkers in subsequent reactions, it is necessary to have perfectly blunt ended ds cDNA. Add the following reagents to the dried ds cDNA pellet:

dH$_2$O	6 µl
5 × TA buffer (166 mM Tris acetate, pH 7.8; 333 mM K acetate; 50 mM Mg acetate; 2.5 mM DTT; 500 µg/ml BSA)	8 µl
RNase A (2 mg/ml)	2 µl
RNase H (1000 U/ml)	2 µl
1 mM β-NAD	2 µl
E. coli DNA ligase (5 U/µl)	2 µl
3 mM dNTPs	(4 µl of each) 16 µl
T$_4$ DNA polymerase (200 U/µl)	2 µl

Incubate at 37°C for 30 minutes.

Increase the volume to 100 µl with TE buffer and add 10 µg of tRNA to serve as a carrier in subsequent reactions. Phenol and ether extract, precipitate and lyophilize.

Methylation Reaction

In this reaction, internal Eco RI sites in the ds cDNA are blocked. This prevents them from being digested by Eco RI following the ligation of Eco RI linkers.

Redissolve the DNA in 25 µl of methylase buffer (100 mM Tris, pH 8.0; 10 mM EDTA; 5 mM DTT). Add 1 µl of 120 mM S-adenosyl methionine and 1 µl (~20 U) of Eco RI methylase. Incubate at 37°C for 1 hour.

Increase the volume to 100 µl with TE buffer. Phenol and ether extract and lyophilize.

Linker ligation

Redissolve pellet in the following:

dH$_2$O	19.5 µl
10 × ligase buffer	2.5 µl
Phosphorylated Eco RI linkers (0.5 µg/l)	2 µl
DNA ligase	1 µl

Incubate at 14°C overnight.
Then heat inactivate ligase at 68°C for 5 minutes.

Eco RI digest

Increase the volume of the ligation reaction to 82 µl with dH$_2$O. Add 10 µl of 10 × Eco RI buffer and 8 µl (~180 units) of Eco RI. Digest for 4–5 hours at 37°C.

Size selection
In a siliconized Pasteur pipet, prepare a column of Sepharose CL 4B. The Sepharose should be pre-equilibrated in column buffer (600 mM NaCl; 10 mM Tris HCl, pH 8.0; 1 mM EDTA; 0.1% sarcosyl).

Load the sample onto the column. Collect 50 μl aliquots. Monitor the elution of the DNA with a Geiger counter.

At this point, individual column fractions, along with aliquots from previous reaction, may be electrophoresed on a 1% alkaline agarose gel to check the size distribution of the cDNAs. We pool column fractions containing material > 500 bp in length. They are precipitated directly, without further salt addition, by adding 2–3 volumes of 95% ethanol. 5 μg of tRNA should be added as carrier.

Ligation of cDNA to λgt arms and in vitro packaging
High quality phosphatased λgt 10 and 11 arms are readily available commercially (Vector Cloning Systems). We recommend using the ligation protocol supplied by the manufacturer which gives an excellent yield of recombinant clones (generally in excess of 5×10^6 recombinants/μg of ds cDNA). It should be remembered, however, that each batch of cDNA should be test-ligated with a set amount of phosphatased λgt arms to determine the maximal ratio of phage:cDNA prior to any large-scale ligation. Small differences in this ratio can make dramatic differences in the yield of recombinants.

In vitro packaging of ligation reactions may be performed as described in the protocols supplied with commercial packaging extracts. Commercially available extracts are highly efficient, with yields 5–50-fold in excess of what is commonly attained with the most commonly utilized procedures.

PREPARATION OF GENOMIC LIBRARIES

Preparation of complete genomic libraries in bacteriophage or cosmids, while now a standard technique with proven success in many laboratories, is not a trivial task. The first question to be answered as one ventures into this area, therefore, is whether the *de novo* construction of a library is needed, or whether one of the available amplified libraries, which have been used to clone many different sequences, would suffice. In particular, the Maniatis phage library[4] has been well maintained, is 'deep' in the sense of containing a large number of independent clones, even in its amplified form, and has been the source of many useful genomic clones. Cosmid libraries are much more difficult to store, since their presence in bacterial colony form allows the possibility of contamination with other flora. However, protocols for storage of cosmid libraries exist (see above description of storage of plasmid cDNA libraries, and reference[5]), and if one is fortunate enough to have access to one, it may be preferable to starting from scratch.

If a library is unavailable, or if one wishes to clone the gene from a mutant individual, then several considerations are appropriate before deciding on a library construction strategy. (A good discussion of many of these issues is given in Ch. 9 of the Maniatis Manual.)[1] An important formula to consider gives the number of independent clones N which must be present in the library in order to have a probability P of a single copy sequence being present:

$$N = \frac{\ln(1-P)}{\ln(1-f)}$$

Here f is the fractional proportion of the genome in a recombinant, which is equal to the average insert size of a clone n divided by the genome size (3×10^9 bp for the haploid human). Since f is small, this can be simplified to

$$N = \frac{3 \times 10^9}{n} \ln(1-P)$$

which makes it clear that the larger the insert, the fewer clones must be generated. This is a major impetus for constructing phage libraries with large inserts (15–20 kb) or using cosmids (inserts 30–45 kb).

If one is attempting to clone a gene for which the restriction map is known, a considerable simplification of the task can often be achieved. This can be done by:

1. identifying a genomic restriction fragment containing the region of interest, preferably one which is either larger or smaller than the majority of fragments generated by this enzyme,
2. digesting genomic DNA to completion with this enzyme,
3. selecting the desired size fragments using a sucrose gradient (see below), and
4. ligating these fragments into an appropriate vector.

The enrichment provided can reduce the number of clones which must be generated and screened by a factor of 10 or more. For example, the human insulin gene resides on a 16 kb Eco RI fragment, whereas most Eco RI fragments are considerably smaller. Similarly, we have taken advantage of the presence of a very large 39 kb Kpn I fragment containing the β-globin cluster to facilitate the cosmid cloning of this cluster in intact form in the hereditary persistence of fetal hemoglobin syndromes.[6]

Another option to be considered at the outset is the use of special vectors and hosts which allow screening through recombinational methods, the so-called πvx system.[7] This approach can also be applied to cosmids.[8] While considerable savings in time can be achieved in the screening steps through this route, much care must be taken to include proper controls, as anomalous events also can give rise to plaques in the final library. A full discussion of this approach is outside the scope of this review.

Phage λ libraries

The most useful λ vectors which accommodate 15–20 kb inserts are given in Figure 5.1. Currently the most straightforward strategy is to use a Bam HI cloning vector such as λCh30[9] or EMBL3A[10] into which is ligated a partial Mbo I digest of genomic DNA. This should generate an essentially random library, since Mbo I sites occur approximately every 256 bp. If a particular restriction fragment is to be cloned, one can consult published listings[1,11] for the appropriate vector.

The vector DNA can be prepared as described above. The vector must then be cut and the middle stuffer segment, if one is present, removed. While gel separation can be used, we have found much greater success in subsequent ligatability using sucrose gradient purification.

100 μg of vector DNA is restricted and a small amount tested on an analytical gel to confirm cutting. The cohesive ends of the vector are then annealed by incubating in 10 mM MgCl$_2$ at 42°C for one hour. A 10 to 40% sucrose gradient is then prepared in an SW27 or SW28 ultracentrifuge tube using ultrapure sucrose in a buffer of 10 mM Tris, pH 7.4, 100 mM NaCl, and 6 mM EDTA. A convenient way to prepare the gradient is to make a solution of 25% sucrose in the above buffer and place this at −20°C until frozen solid. The tube is then removed and allowed to thaw at room temperature in a rack which allows free flow of air around the tube. As thawing occurs the heavier liquid falls to the bottom, and an almost exactly 10–40% gradient is formed (this can be easily checked using a refractometer).

The sucrose gradient is loaded with the cut vector DNA and centrifuged at 23 000 rpm for 20–24 hours in a Beckman SW27 or 28 rotor. Fractions are analyzed and pooled as described in the Maniatis cloning manual, except that we find the desired fractions can be precipitated directly from the sucrose buffer without dialysis. After resuspension in TE buffer, the DNA should be checked for self-ligatability.

VECTORS	MAP	VECTOR SIZE (kb)	CLONING ENZYME	STUFFER (kb)	INSERT RANGE (kb)
λCh4A	RI RI RI 19.9 ⎯ 6.6, 7.8, 11.0	45.4	EcoRI	6.6 +7.8	7.1–20.1
λCh21A	1.9 21.7 RI/ 18.3 Hind III	41.9	EcoRI Hind III	–	0–9.1
λCh30	Bam Bam Bam 22.7 ⎯⎯ 7.4 ⎯ 7.4 ⎯ 9.3	46.8	Bam HI	7.4 +7.4	6.1–19.1
λEMBL3A	Bam Bam Sal RI RI Sal 20.3 ⎯ 13.7 ⎯ 9.2	43.2	EcoRI Bam HI	13.7	8.6–21.6

Fig. 5.1 Skeletal maps and vital statistics of the most commonly used λ cloning vectors. Those marked with an 'A' (all except λCh30) bear amber mutations in two or more λ genes, which requires that they be grown on bacterial hosts which carry a suppressor gene. For more detailed maps, see Maniatis et al (for the Charon vectors)[1] or Frischauf et al (for EMBL3A).[10]

Preparation of the genomic DNA from tissue culture cells or peripheral blood should be carefully carried out using the Blin & Stafford protocol[12] (see Ch. 2). It is important to avoid any pipetting as this will shear the DNA. Note that this protocol avoids the use of ethanol precipitations — intact genomic DNA may be very difficult or impossible to get back into solution after precipitation. It is *absolutely essential* to check the genomic DNA on a 0.3% or 0.4% gel (poured over a 1% bed) using intact λDNA as a size marker. If more than 10% of the DNA is less than 50 kb in size, efficiencies in library generation will suffer.

Partial Mbo I or Sau 3A (both of which produce overhanging ends which can be ligated to Bam HI vector ends) digestion of the genomic DNA can be performed exactly as described in Chapter 2 Maniatis Manual. It is important to include, as a control, DNA which is incubated in the buffer alone without any restriction enzyme, to be certain that the decrease in the size of DNA with incubation is not just due to activation of nuclease in the DNA preparation from the magnesium in the buffer. Any evidence of nuclease contamination should lead to re-extraction with phenol. Sucrose gradient centrifugation (with the gradients prepared as above, except that the NaCl can be omitted) generally gives DNA of higher ligatability than gel electrophoretic purification. Once precipitated and resuspended, the size-selected DNA should be tested for self-ligatability by ligating 1 μg in 10 μl (see protocol earlier in this chapter) and analyzing the result on a 0.3–0.4% gel, with unligated DNA run in the adjacent lane as a control.

A decision must now be made about phosphatasing*. If the vector has a large stuffer and the separation of arms has been successful, phosphatasing the vector should not be needed, since the arms ligated to each other without an insert will be too small to be packageable (packaging range is 35–51 kb, though in the 35–40 kb range the efficiency falls off rather precipitously). Whether to phosphatase the insert DNA is more problematic. The purpose of doing this is to reduce the possibility of generating clones with two insert fragments, which can be highly misleading if one assumes that the two inserts represent contiguous sequences in the genome. However, two other factors act to reduce the occurrence of these multiple insert clones: one usually uses a vector to insert ratio of about 4:1 in the ligation, and if the insert is well size-selected, a double insert ligation will usually be too large to package. Neither of these factors will eliminate the problem, however, and if the cloning situation is such that a tandem insert would lead to major difficulties, phosphatasing the insert DNA is recommended. Use of calf intestinal phosphatase (CIP) and the heat killing and phenol extraction recommended in the Maniatis Manual (p. 133) is usually successful, and the Sephadex G-50 step can be omitted if one uses a CIP such as Boehringer Mannheim molecular biology grade, which is not suspended in ammonium sulfate. For unclear reasons, however, phosphatasing the insert will usually lead

---

* Editor's note: DNA ligase requires that the 5' sugar positions of at least one of the ligating fragments have phosphates on them. Phosphatase removes the 5' phosphate. Thus, if the inserts are phosphatased, but the vector arms are not, the inserts can ligate with vector arms but not other insert fragments.

to a 4 to 8-fold reduction in efficiency of plaque formation of the ligated DNA. Therefore, if a tandem insert ligation would not be a problem (as is often the case when the genomic map of the region to be cloned has already been established and one is cloning a mutant gene), phosphatasing is probably better omitted.

A test ligation of vector and insert, together with a control of vector alone, should now be carried out. The DNA concentration in the ligations should be at least 0.2–0.3 µg/µl. We find it helpful to remove 1 µl of the ligation mixture just prior to adding the ligase enzyme and to set this aside. The unligated and ligated DNA should then be run side by side on a 0.3–0.4% gel to assess whether the size of the DNA has significantly moved upward after ligation.

Test-ligated vector alone and vector + insert should be packaged and plated on the appropriate bacterial host which has been grown in 0.4% maltose to promote expression of the phage λ receptor. A parallel reaction containing control intact λ DNA should always be included in the packaging experiment and should yield at least 1×10^8 pfu/µg if the extracts are good. A good library will give $> 10^6$ pfu/µg of ligated DNA, though one can still make a complete library with less efficient yields. The self-ligated vector should give 10% or less of the number of plaques as vector + insert.

Before proceeding to scale-up and large scale plating and screening, we find it useful to screen a plate of approximately 1000 plaques using nick-translated total human DNA as a probe; nearly all phage plaques with human inserts of 15–20 kb should hybridize because of the presence of repetitive sequences. It is also a good idea to pick 6–10 plaques and prepare DNA using a 'minilysate' protocol such as that given above. Digestion of the DNA can be performed to demonstrate that each plaque contains a different insert and that the vector arms are intact.

Cosmid libraries
Cosmids have the advantage of the largest insert size (up to 45 kb) of any cloning vector. Their name derives from the presence of the cos (cohesive end) signal of phage λ in the plasmid vector used in cloning. The vector also contains an antibiotic resistance gene and a bacterial origin of replication, so that it is grown as a plasmid. However, when genomic DNA is ligated into the vector so that a distance of 35–51 kb separates two cos sites, the DNA can then be packaged *in vitro* and efficiently transfected into host cells. Therein lies the strength of the cosmid approach: such large molecules would be extremely inefficient to transfer into *E. coli* using other transfection procedures.

Cosmid cloning is, however, more difficult than cloning in phage λ, and more care and optimization of each step is required to generate a complete library. The techniques described in Maniatis (p. 300–307),[1] which rely upon the method of Ish-Horowicz & Burke[13] are basically reliable. To this description we would add the following comments:

1. pJB8 is the most widely used vector, although others with selectable markers are available.[14] pJB8 is usually used to clone 30–45 kb sized genomic

DNA which has been partially digested with Mbo I or Sau 3A.

2. It is absolutely essential that the starting genomic DNA be greater than 100 kb in size before digestion. If it is not, there is no point in beginning the procedure.

3. If self-ligation of the vector is not prevented by some technique, many cosmid clones will have tandem vector inserts which will reduce the average insert size significantly. The Ish-Horowicz & Burke solution to this problem is clever, effective, and highly recommended. This leaves the cloning site of the vector unphosphatased. The genomic insert DNA can then still be phosphatased to prevent tandem ligation and generation of clones with multiple inserts. A price will be paid in efficiency, however. As in phage cloning, the particular situation will dictate how important phosphatasing the insert may be. A good discussion of the effects of the various options on cloning efficiency can be found in Grosveld et al.[5]

4. It is important to use a bacterial host which is recA$^-$, as the large size of cosmids makes them prone to deletion and rearrangement in recA$^+$ hosts. HB101 is a good cosmid host, and its recA$^-$ phenotype should be frequently checked by exposing a plate just after streaking to a hand-held UV light for 30 seconds. This will prevent growth of recA$^-$ but not recA$^+$ strains.

5. The major problem with cosmid library generation is efficiency. Since a complete library should contain at least 300 000 independent clones, it is desirable to have an efficiency of at least 50 000 to 100 000 clones per μg of ligated DNA. This requires high quality vector and insert DNA, and packaging extracts of excellent quality (preferably 4×10^8 pfu/μg of intact λDNA or better).

6. For unclear reasons, there is a great deal of variability from one overnight culture to the next in the plating efficiency of packaged cosmids. This is not just a function of the bacterial strain or the cell density at the time of harvesting. It is therefore a good idea to grow several 3 ml host cultures, spin down the bacteria at 3000 rpm in a table top centrifuge for 5 minutes, pour off the broth and resuspend the bacteria in 1/2 volume of 10 mM $MgCl_2$. The host bacteria will be stable for up to one week, while each overnight is tested for plating efficiency with a small amount of packaged cosmids. The best one can then be chosen for large scale plating.

7. Screening of cosmids is analogous to screening of plasmids, which is covered in Chapter 6. However, because of the low copy number of cosmids relative to plasmids, the hybridization signals can be faint. An approximately 4-fold improvement in signal can be achieved by placing the nitrocellulose filters bearing the cosmid clones (just after pulling these from the plates) on fresh LB-chloramphenicol (170 μ/ml) plates, with colonies up, and leaving these at 37°C overnight prior to denaturation.

REFERENCES

1. Maniatis T, Fritsch E F, Sambrook J. Molecular cloning. A laboratory manual. Cold Spring Harbor Laboratory, 1982.
2. Young R A, Davis R W. Efficient isolation of genes by using antibody probes. Proc Natl Acad Sci USA 1983; 80: 1194.
3. Prochownik E V, Markham A F, Orkin S H. Isolation of a cDNA clone for human antithrombin III. J Biol Chem 1983; 258: 8389.
4. Maniatis T, Mardison R C, Lacy E et al. The isolation of structural genes from libraries of eukaryotic DNA. Cell 1978; 15: 687.
5. Grosveld F G, Dahl H M, deBoer E, Flavell R A. Isolation of β-globin related genes from a human cosmid library. Gene 1981; 13: 227.
6. Collins F S, Stoeckert C J, Serjeant G R, Forget B G, Weissman S M. $^G\gamma\beta^+$ Hereditary persistence of fetal hemoglobin: cosmid cloning and identification of a specific mutation 5' to the ^{G} γgene. Proc Natl Acad Sci USA 1984; 81: 6894.
7. Seed B. Purification of genomic sequences from bacteriophage libraries by recombination and selection *in vivo*. Nucleic Acids Res 1983; 11: 2427.
8. Poustka A, Rachwitz M R, Frischauf A M, Hohn B, Lehrach H. Selective isolation of cosmids clones by homologous recombination in *E. Coli*. Proc Natl Acad Sci USA 1984; 81: 4129.
9. Rimm D L, Morness D, Kucera J, Blattner F R. Construction of coliphage lambda charon vectors with BamHI cloning sites. Gene 1980; 12: 301.
10. Frischauf A M, Lehrach H, Poustka A, Murray N. Lambda replacement vectors carrying polylinker sequences. J Mol Biol 1983; 170: 827
11. Williams B G, Blattner F R. In Setlow J K, Hollander P, eds. Genetic Engineering. vol 2. New York: Plenum, 1980: p. 201.
12. Blin N, Stafford D W. Isolation of high molecular weight DNA. Nucleic Acids Res 1976; 3: 2303.
13. Ish-Horowicz D, Burke J F. Rapid and efficient cosmid cloning. Nucleic Acids Res 1981; 9: 2989.
14. Lau Y F, Kan Y W. Versatile cosmid vectors for the isolation, expression, and rescue of gene sequences: studies with the human β-globin gene cluster. Proc Natl Acad Sci USA 1983; 80: 5225.

6
Molecular cloning of genes: use of hybridization probe screening methods
J. A. Kant

In this chapter we will examine how a 'cloned' copy of a gene or its mRNA product is obtained by using radiolabeled nucleic acid probes (Ch. 4) to screen vector libraries containing cellular genomic fragments or cDNAs (Ch. 5). Probe screening is a comparatively simple procedure, as you will see if you have succeeded in the more difficult task of constructing such vector libraries and isolating or synthesizing nucleic acid fragments for use as probes. Even should you 'clone by phone' obtaining, say, woolly mammoth cDNA and genomic libraries from one associate and a human globin probe from another, you should have little difficulty in obtaining a purified woolly mammoth globin gene and cDNA within a month. A general discussion of hybridization probe screening is presented first, followed by protocols for screening both bacterial plasmid and bacteriophage vector libraries with examples from the author's experience. These procedures should be generally applicable to other systems. The chapter concludes with a discussion of refinements and other applications of hybridization probe screening that may be useful for particular situations.

A. AN OVERVIEW OF HYBRIDIZATION PROBE SCREENING TECHNIQUES

Hybridization probe screening techniques were first developed by Grunstein & Hogness[1] for application to bacteria containing plasmids and by Benton & Davis[2] for screening of bacteriophage libraries. The techniques are based on the re-annealing of single stranded radiolabeled nucleic acid probes to complementary regions of denatured genomic DNA or cDNA in the vector library, much as occurs with Southern and Northern blots (Chs. 2 and 3). Successful library screening requires ready detection of individual members containing complementary sequences and the isolation of these free from other members of the library.

The cDNA or genomic DNA fragments in individual members of the vector library must first be *amplified* to provide sufficient copies for timely hybridization of the labeled probe. The vector library is first grown on a suitable medium that may require antibiotics if the library is in a plasmid which contains an antibiotic resistance gene. As each unique plasmid-containing bacterium replicates under selective conditions to form a colony, there is also replication of plasmid DNA amplifying many times an inserted cDNA or genomic DNA fragment. Similarly, bacteriophage infection, replication and lysis of a single bacterium, with subsequent reinfection of adjacent bacteria by progeny phage, generates expanding clear zones in the bacterial lawn, called plaques. The plaques are free of bacteria but contain thousands of identical phage, each with a copy of any new sequences inserted into the DNA of the initial infecting phages during construction of the library.

Amplification of the library is generally followed by a *templated transfer* of amplified material to a support medium such as nitrocellulose filter paper for further processing. The original amplified sample from which the transfer replica is made is preserved as a 'master' from which additional replicates can be made or samples taken for further purification. Large numbers of bacteriophage ($2-10 \times 10^4$/150 mm plate) or bacterial colonies (up to 10^5 or more per 82 mm nitrocellulose filter) can be examined by these methods, making library screening a convenient experimental procedure which can be performed on relatively few plates.

The amplified library DNAs on nitrocellulose are treated in the presence of a strong base to *denature DNA into single stranded forms*. The paper is then neutralized with a strong buffer, still preserving much of the DNA in single stranded form. Baking dry filters under vacuum assures strong fixation of DNA to the nitrocellulose. The filters are pre-hybridized with solutions designed to reduce non-specific background binding of nucleic acid probes, and then *hybridized under appropriate conditions to denatured (single stranded) radiolabeled nucleic acid probes*. The probe binds to regions of the paper containing amplified copies of vector DNA with complementary sequences. Non-specific binding of radiolabeled probe to other sequences is minimized by washing the papers at temperatures and with solutions that favor binding of the radiolabeled probes only to highly homologous or identical sequences. *Autoradiography of washed filters allows the selection of individual vector members containing the sequences of interest.* These bacteria or phage containing a single genomic or cDNA fragment can then be grown as a pure population in large quantity to yield DNA for restriction endonuclease mapping, nucleic acid sequencing, cell transfection and a host of other experiments.

B. PROBE SCREENING OF PLASMID LIBRARIES BY COLONY HYBRIDIZATION

General materials and solutions:
Plasmid cDNA library — see Chapter 5.
Nitrocellulose filter circles (82 or 137 mm for 100 or 150 mm petri dishes).
Whatman 3MM or other blotting paper.
Bacterial growth medium (e.g. Luria-Bertani (LB) with magnesium — 1% tryptone, 0.5% yeast extract, 0.01 mol/l NaCl, 0.01 mol/l $MgCl_2$).
Appropriate antibiotics for plasmid library growth (e.g. tetracycline 20 μg/ml; ampicillin 30 μg/ml).

LB-agar plates (LB medium + 1.5 g/dl agar autoclaved, appropriate sterile antibiotics added, then plates poured).
SET — 0.15 mol/l NaCl, 1 0.001 mol/l EDTA, 0.03 mol/l Tris HCl, pH 8.0.
Denhardt's solution — 0.02 g/dl% each of Ficoll, bovine serum albumin, and polyvinylpyrrolidone dissolved in water.
Denaturing, neutralizing and wash solutions — described below.
Formamide (if used) — deionized with ion-exchange resin AG501-X8, Bio-Rad, or similar and stored frozen.
Sodium dodecyl sulfate (SDS).
^{32}P-labeled nucleic acid probe — see Chapter 4.

1. Preparation
One can set up filters for colony hybridization depending on whether relatively few (< 500) or many colonies are to be screened.[3] For non-abundant species, such as mRNAs for rare polypeptides, it is necessary to examine large numbers of recombinant clones from your library to identify one or two with the appropriate sequences.

Limited numbers of colonies can be consolidated in a defined pattern (say a 10 × 10 grid) by transfer with sterile toothpicks onto a round 82 mm nitrocellulose filter which has been placed on a fresh LB agar-antibiotic plate then grown overnight at 37°C. Colonies can also first be toothpicked to and grown in 0.1–0.2 ml of media in 96 well microtiter plates and transferred to 82 mm nitrocellulose filters 48 at a time using a custom fabricated 6 × 8 replica plating device consisting of a metal base with 48 one inch metal tines 2–3 mm in diameter arranged to fit the microtiter plate. The tines are dipped in 70% ethanol and flamed between platings to prevent cross-contamination.

Larger numbers of colonies (10^4–10^5/plate) are applied as bacterial suspensions in media spread to cover the inner 90% of a presterilized nitrocellulose filter laid on a fresh plate (0.4 ml for 82 mm filters; 0.8 ml for 137 mm filters). The plates are inverted and incubated at 37°C until colonies are barely visible (approximately 0.1 mm). Incubation at lower temperatures (30–32°C) may facilitate control of growth to the desired size. This 'master' filter is then lifted carefully from the plate using flat-ended forceps and placed on sterilized Whatman 3MM paper, colonies facing up. A second nitrocellulose filter numbered to correspond to the master is pre-wetted by placing on a fresh plate, numbered side down. This filter is then placed number side down atop the master and pressed gently with a flat object (we use Hanahan's[3] suggestion of a flat block covered with velvet). Before the filters are separated, it is essential to key the replica to the master by puncturing a characteristic pattern of holes through both using a 20 gauge needle (e.g. 2 holes 20–30 mm apart at one edge and 1 hole at the opposite edge of the filter). The replica is placed back on its plate, colony side up and grown at 37°C until colonies are 0.5–1.0 mm in size. Additional replicas (up to four) can usually be made from the master, at which time the master must be reincubated to regenerate colonies. The master filter can be stored for several weeks on a plate wrapped in parafilm (be careful that moisture does not get on the filter and cause smearing of the colonies). For long-term storage, a sandwich replica of two filters containing small (0.1 mm) colonies can be frozen at −80°C.

2. Colony lysis, denaturation and neutralization of DNA

The simplest way to lyse colonies and denature DNA is to lay the nitrocellulose filter replica, colony side up, on 3MM paper saturated with 0.5 mol/l NaOH for 2–3 minutes and repeat the treatment. Neutralization of the filter follows for 5 minutes on paper saturated with 1 mol/l Tris HCl, pH 7.5, then on paper saturated by 1.5 mol/l NaCl–0.5 mol/l Tris HCl, pH 7.5. Some investigators first place the filter on 3MM paper saturated with 10 g/dl SDS for 3 minutes prior to denaturation and neutralization.[4] This author has also successfully employed a more laborious protocol involving 1 minute incubations of filters, 3 changes each, on successive papers saturated with solutions of:
  a. 1 mg/ml lysozyme in 25 g/dl sucrose, 0.05 mol/l Tris HCl, pH 8.0
  b. 0.2 g/dl Triton X-100, 0.5 mol/l NaOH
  c. 0.5 mol/l NaOH
  d. 1 mol/l Tris HCl, pH 7.5 and
  e. 0.15 mol/l NaCl, 0.1 mol/l Tris HCl, pH 7.5[5]

3. Dry and bake filters

The filters are then air-dried at room temperature for 30–60 minutes and baked between dry sheets of 3MM paper in a vacuum oven at 80°C for 2–3 hours (or overnight at 60°C).

4. Pre-hybrization

The filters are pre-hybridized in heat-sealed plastic bags for smaller numbers of filters (e.g. 4 to 8) or in polypropylene storage dishes for larger numbers. 4 to 5 ml of $6 \times$ SET pre-hybridization mix containing $5 \times$ Denhardt's solution, 0.5 g/dl SDS and 100 μg/ml of sheared, denatured salmon sperm (or other) carrier nucleic acid is added for each filter and the group incubated 2–6 hours at 68°C. The designation $6 \times$ or $n \times$ preceding a solution is a convention used throughout the chapter, designating a 6- or n-flood concentrated solution (e.g. $6 \times$ SET is 0.9 mol/l NaCl, 0.006 mol/l EDTA, 0.18 mol/l Tris HCl, pH 8.0).

5. Hybridization

The pre-hybridization fluid is decanted from the filters. To an equal volume of hybridization solution ($6 \times$ SET, $1 \times$ Denhardt's, 0.1 g/dl SDS), add denatured single stranded radiolabeled probe at approximately 0.5–1.0×10^6 (2.2–4.5×10^{-7} B_q) dpm/filter (specific activity 1–5×10^8 dpm/μg (4.5–22.5×10^{-5} B_q) probe). The probe should be in a low ionic strength solution (such as TE buffer) after labeling and denatured before adding to the hybridization solution by 1) heating in a microcentrifuge tube for 5–10 minutes in boiling water or 2 by adding to probe 0.1 volume of 1 mol/l NaOH with incubation for 2) minutes at room temperature followed by neutralization with 0.1 volume Tris HCl, pH 7.5 and 0.1 volume 1 mol/l HCl. Hybridization is carried out overnight in this solution at 68°C or for 2 days at 37–42°C using the same solution with 50% formamide, a compound which lowers the melting temperature of double stranded DNA. The formamide method is slower but is less strenuous on the filters and reduces problems of overnight evaporation if the hybridization is being done in a 68°C water bath. Constant agitation of the hybridization mix is usually not necessary but may help reduce background binding of probe.

6. Washing and autoradiography of filters

After hybridization, wash the filters 2 or 3 times for 5–10 minutes at room temperature to dilute any remaining hybridization fluid. A non-stringent solution such as 2 × SET, 0.1 g/dl SDS at 10 ml/filter is satisfactory. Thereafter several 30 minute washes with the same solution at 68°C should render the filters quiet when examined with a ^{32}P-sensitive radiation survey meter. If increased stringency of hybridization is desired, the filters are washed further with solutions of lower ionic strength at the same or reduced temperature (e.g. 0.1 × SET, 0.1 g/dl SDS at 50°C for 30 minutes is a relatively stringent wash).

After drying, the filters are taped to a solid support (paper or old X-ray film). Up to 20 82 mm filters can be autoradiographed simultaneously in a 14" × 17" standard X-ray cassette. The support should contain several areas marked with a distinctive pattern of radioactive ink, which allow precise localization on the autoradiographed filters of the holes made to key the filters to the 'master' template in Step 1 above. Radioactive ink is made by adding unused isotope or probe remaining from previous experiments to India ink in a microcentrifuge tube. These marks are not necessary for low-density replica plated filters. The filters are placed in a standard metal or cardboard X-ray cassette adjacent to a piece of X-ray film (e.g. Kodak X-omat-AR), and 1 or 2 intensifying screens and exposed overnight at −70°C. The cassette is thawed and the film developed the next morning. Dark spots indicate areas of hybridization with your probe (see Fig. 6.1). Those hybridizing most darkly may be clones with the longest inserts homologous to your probe or those most homologous if you are detecting related but divergent species. If desired, the stringency of wash solutions, as well as hybridization conditions, can be varied to minimize or maximize differences.

7. Further purification of hybridizing clones

If you have screened at low density, the bacterial clone(s) containing plasmids with sequences complementary to your probe will already be purified on a master plate or in a microtiter plate. However, if you have screened at high density, the exact colony contributing the signal on the master filter will probably not be clear. You will need to take a scraping of colonies from this region using a replica of the master, and dilute them in media such that several hundred colonies grow up when spread on a new nitrocellulose filter. At least several of the colonies which grow up should be the clone of interest and be separable as a pure colony by going through the probe hybridization screening steps already outlined a second time.

Example

An example of colony hybridization screening at low density is shown in Figure 6.1. Here cDNAs for rat α (a), β (b) and γ (g) fibrinogen chains were used as hybridization probes to 3 replica filters (6 rows, 8 columns) of cDNA clones from a human liver library that had been previously selected as putative fibrinogen candidates by hybridization to a mixture of the 3 probes. It is clear that hybridizing clones bind only 1 of the 3 probes. The 'ghost' effect seen with many hybridizing clones represents smearing of the lysed bacterial contents because filters were lifted slowly, always from the same side, during treatments with lysing, denaturing and neutralizing solutions.

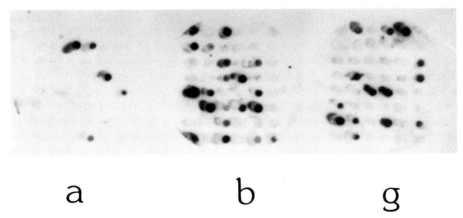

Fig. 6.1 Colony hybridization using rat fibrinogen α (a), β (b) and γ (g) cDNAs to identify the corresponding human cDNAs. Experimental details are provided in the text.

C. HYBRIDIZATION PROBE SCREENING OF BACTERIOPHAGE GENOMIC LIBRARIES

General materials and solutions

Suitable host bacterial strain for bacteriophage (e.g. *E. coli* LE392) — see Chapter 5.
LB medium with magnesium — described in Section B above.
Nitrocellulose filters, 82 mm and 137 mm sizes.
LB-agar plates, 100 and 150 mm, described in Section B above, without antibiotics.
SM buffer (0.01 mol/l MgSO$_4$, 0.1 mol/l NaCl, 0.05 mol/l Tris HCl, pH 7.5, 0.02 g/dl gelatin), sterile.
Chloroform.
SET — described in Section B above.
Denhardt's solution — described in Section B above.
Denaturing, neutralizing and wash solutions — described below.
Radiolabeled nucleic acid probe — see Chapter 4.

1. Titer the library

The day prior to plating your phage clones for screening, the titer of the phage stock should be checked so you will know exactly what volume to use to obtain the desired number of phage plaques. Approximately 8×10^5 clones must be screened in a genomic library of average 17 kb insert size to have a 99% probability of detecting a given unique sequence.[4] Generally $3-10 \times 10^5$ phage are screened. A fresh overnight culture of host bacteria is also grown for use in the definitive screening experiment the following day.

Use 10-fold dilutions of phage over 3 to 4 logs to determine the titer, adding the phage in 10 μl of SM buffer to 0.2 ml of mid-log phase bacteria in a 5 or 10 ml test tube. The phage are quite sticky, and it is important to rinse thoroughly but gently the pipet used in adding phage to dilutions of bacteria. Incubate for 10-15 minutes at 37°C to allow phage attachment then add 4 ml of 0.6 g/dl agar or agarose in LB-magnesium media at 50°C. The top agar-bacteria-phage mixture is then poured quickly before cooling onto a room temperature LB-agar plate and swirled into a uniform thin coating. After 5 minutes to harden, the plates are incubated inverted overnight and the titer calculated from the number of plaques observed the following morning: titer (plaques/ml) = # plaques/plate × 100 × dilution factor. The easiest way to determine an accurate titer is to count plates with 50-300 colonies. Most phage stocks have titers of 10^9 to 10^{12} plaques/ml.

2. Bacteriophage infection and growth

Infection is similar to the titering process. The number of phage you wish to screen are added from a dilution of phage stock in SM buffer to mid-log phase host bacteria in LB-magnesium media, using about 0.7 ml of bacteria per 150 mm plate to be poured. Thus, you would add 100 μl of a 10^{-4} dilution of a 10^{11} plaques/ml phage stock to 14 ml of bacteria to screen one million phage on 20 plates at a convenient screening density of 50 000 plaques/plate. Lesser or greater numbers of phage can be screened as discussed. After a 10 minute incubation at 37°C, the bacteria and phage are added to a volume of 0.6 g/dl top agarose at 50°C (see above) sufficient to pour 20 plates (200 ml). Agarose is preferred to agar because it is less likely to strip off when the contents of plaques are transferred later to filters and gives lower levels of background hybridization. Using a pipet, 10 ml aliquots are spread evenly onto room temperature 150 mm LB-agar plates and swirled to form a thin layer of top agarose. Allow the plates to harden for 5-10 minutes then grow inverted overnight at 37°C.

3. Transfer of phage, denaturation and neutralization

The plates should be a sea of small plaques the following morning, with very little or no bacterial lawn evident. 20 137 mm nitrocellulose filters are numbered 1-20 with a ball point pen and placed number up on the top agar of the respective plates for 30-60 seconds. The nitrocellulose should wet completely over 5-15 seconds, and we find it is best to lay half of the filter on the plate and then allow the wetting action to proceed in one direction across the rest of the filter. The orientation of the filter is keyed by poking small holes through the filter into the agar with a 20 gauge needle that has been dipped in India ink. This leaves recognizable ink tracks in the agar of the primary plate that are useful later for keying the plate to autoradiograms. We have encountered no problems with the loss of top agarose removing filters from such plates at room temperature, though some investigators recommend chilling the plates at 4°C for an hour before 'pulling' the nitrocellulose filters.[4]

After 1 minute the filters are gently removed by lifting up a corner with the help of flat forceps and peeling the filter off the plate. The filter is immediately dropped into a denaturing solution of 0.5 mol/l NaOH, 1.5 mol/l NaCl for 30–60 seconds, then placed in a neutralizing solution of 1.5 mol/l NaCl, 0.5 mol/l Tris HCl, pH 7.5 for 3–5 minutes. We have found it useful to pull 1 set of replicate filters for this first round of screening to confirm hybridization signals. A second set of nitrocellulose filters (1B–20B) is applied after removal of the first set, keyed to the plates using the same tracks as the first set of filters, and allowed to incubate 3–5 minutes at room temperature. These filters are then removed, denatured and neutralized as the others.

4. Drying and baking

After neutralization the filters are allowed to dry 30–60 minutes at room temperature before baking 2–3 hours at 80°C in a vacuum oven.

5. Pre-hybridization, hybridization and washing of filters

These steps follow for the most part the procedures described above for colony hybridization screening. We generally add $1-2 \times 10^6$ cpm of probe per filter, and wash twice with $6 \times$ SET, 0.1 g/dl SDS for 30 minutes at 68°C, then twice with $1 \times$ SET, 0.1 g/dl SDS for 30 minutes at 68°C. For reproducibility we prewarm the wash solutions.

6. Autoradiography and further purification of phage clones

After autoradiography, as described for colony hybridization, the keying marks on the filters are used to mark the autoradiographs. At this point compare replicate filters to ensure hybridization signals appear on both filters pulled from the same plate. The keying marks are used to orient the original phage plates over positive hybridization signals. A plug of top agar encompassing the hybridization signal is removed with the wide end of a Pasteur pipet and placed into SM buffer with a drop of chloroform. After 1 hour or so at 4°C this suspension of phage can be used for further screening by repeating the above screening procedures.

By plating several dilutions of the primary isolates, you may be fortunate to find an isolated plaque on the second round of screening, allowing purification within 2 weeks. A single batch of labeled probe is usually sufficient for an entire procedure involving several rounds of screening because less probe is required on subsequent rounds as the number of filters and plaques/plate declines during purification. If probe is particularly scarce, one can even freeze at -70°C the

96 MOLECULAR GENETICS

hybridization mixture from the first round of filters screened and use thawed aliquots (which contain residual single stranded probe) for subsequent rounds.

Example

Figure 6.2 illustrates the screening of a bacteriophage genomic library to attain plaque-purified phage clones over 3 rounds of screening. Human fibrinogen cDNA probes for the α (a), β (b) and γ (g) chains were used to screen 10^6 plaques

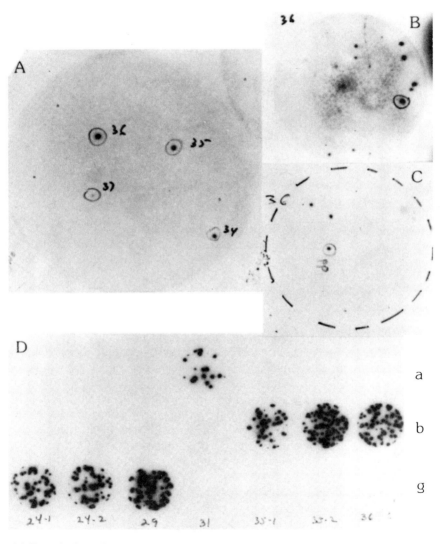

Fig. 6.2 Bacteriophage library screening to isolate human α (a), β (b) and γ (g) fibrinogen genes. Experimental details are provide in the text.

from a bacteriophage genomic library in λ phage Charon 4A. 40 filters were pulled from 20 plates, each containing 50 000 plaques, and hybridized to 3×10^7 cpm of each human cDNA. A representative filter with hybridization signals is shown in panel A. Primary isolate #36 was rescreened with the mixture of 3 probes to yield 8 to 10 secondary isolates (panel B), none clearly discrete enough to pick as a pure clone. Rescreening of a secondary isolate (circled) yielded discrete plaques which hybridized with the mixed probe (pp-panel C). Similar plaque-purified primary isolates hybridized to individual a,b or g probes on small nitrocellulose filters, showed that each isolate contained sequences from a single gene, the β-chain gene in the case of isolate #36.

D. OTHER ALTERNATIVES IN HYBRIDIZATION PROBE SCREENING

Oligonucleotides

Oligonucleotides, short chemically synthesized stretches of DNA, are becoming one of the most common probes used to screen vector libraries (c.f. Chapter 11, Parts One and Two). Even a short peptide sequence from a cellular product, reverse translated into the appropriate mixture of possible oligonucleotides, can rapidly lead to isolation of a full length cDNA, which can in turn be used to determine a full protein sequence and to isolate the gene. Further discussion of oligonucleotide labeling and use are given in Chapters 4 and 11. The short length of oligonucleotides makes their hybridization behaviour strongly dependent on the content of G-C base pairs, which bind more strongly than A-T pairs. This, coupled with the uncertainty of which nucleotide sequence is actually homologous to the gene, makes it difficult to pick conditions which give sensitive and specific hybridization of oligonucleotides to low abundance sequences in complex libraries. Empirical formulae for estimating melting temperatures of oligonucleotides provide useful guides for hybridization temperatures in such screenings.[3,6]

Tetramethylammonium chloride (Me4NCl) hybridization

Uncertainty in oligonucleotide hybridization conditions has been largely abolished by Wood's recent demonstration that hybridization of oligonucleotides in the presence of 3.0 mol/l Me4NCl is dependent only on probe length, and not on G-C content.[7] Me4NCl at this concentration binds to A-T base pairs, raising the melting temperature to that of the G-C pairs. A 4–16 hour low temperature (37°C) pre-hybridization in $6 \times$ SET, $5 \times$ Denhardt's with 100 μ/ml sonicated salmon sperm DNA is followed by a non-stringent 37°C overnight hybridization in the same solution. Filters are then rinsed in $6 \times$ SET, exchanged into Me4NCl, and washed twice for 20 minutes in 3.0 mol/l Me4NCl at 2–4°C below the empirically determined melting temperature for the length of oligonucleotide probe employed.

Screening cDNA libraries for differentially expressed mRNAs

If one is interested in a particular gene or class of genes whose expression can be up (or down) regulated several fold, cDNA probes made from mRNA of such cells in their basal and regulated states can indicate which clones in a cDNA library show such regulation. For example, mammals are known to compensate for states of consumption or degradation of the clotting factor, fibrinogen, by making significant amounts of new fibrinogen mRNA and protein in their livers.[8] To take advantage of this fact in selecting possible cDNAs for rat fibrinogen, we screened replicates of a rat liver cDNA library with single stranded cDNAs made from mRNA of normal and defibrinated rats.[9] The cDNA probes from defibrinated rat mRNA showed stronger hybridization signals with a number of cDNA clones in the liver library, many of which turned out to be fibrinogen when sequenced.

Subtraction hybridization of cDNAs

In this technique single stranded cDNAs prepared from mRNAs of one cell type are generally 'subtracted' by hybridization to mRNA of another cell type. Non-hybridized, theoretically cell-specific, cDNA remaining is separated from hybridized forms by hydroxyapatite chromatography and can be used as a putatively cell-specific probe for screening vector libraries which themselves may reflect whole cell or 'subtracted' cDNAs. This technique was used successfully to isolate the first molecular clones for the T-lymphocyte antigen receptor.[10,11]

Immunoprecipitation of polyribosomes to enrich for low abundance — mRNAs

Success with this technique has been variable. Korman et al[12] elegantly demonstrated the power of the procedure by using a monoclonal antibody directed against the heavy chain of HLA-Dr molecules to immunoprecipitate polysomes from human lymphoblastoid cell lines followed by binding of complexed polysomes to protein A-sepharose columns, elution and purification of the mRNA. The mRNA was used both to make cDNA libraries highly enriched for HLA-Dr sequences and as a source of cDNA probes to screen whole cell cDNA libraries from HLA-Dr positive lymphoblastoid cells.

Hybridization selection and translation

This technique, discussed in greater detail in Chapter 13, is based on dividing cDNA libraries into groups whose DNAs are then amplified, denatured and attached to nitrocellulose or other activated papers to serve as hybridization probes capable of binding mRNAs from cells known to produce a desired product. Of course, the cDNA library used must also have been made from cells which express the mRNA for the product of interest. After washing to remove non-specifically bound RNAs, residual mRNA is eluted, translated *in vitro* and run on gels, with or without immunoprecipitation, to see if anything in the subdivided groups of the cDNA library contains sequences that select the desired mRNA. If the mRNA of interest is selected by DNAs from a group within the library, the procedure is then repeated, subdividing that group into smaller numbers until a single clone whose DNA selects the mRNA is found.

REFERENCES

1. Grunstein M, Hogness D. Colony hybridization: a method for the isolation of cloned DNAs that contain a specific gene. Proc Natl Acad Sci USA 1975;72: 3961–3965.
2. Benton W D, Davis R W. Screening gt recombinant clones by hybridization to single plaques in situ. Science 1977; 196: 180–182.
3. Hanahan D, Meselson M. Plasmid screening at high colony density. Methods Enzymol 1983; 100: 333–342.
4. Maniatis T, Fritsch E F, Sambrook J. Molecular cloning, a laboratory manual. Cold Spring Harbor Laboratory, 1982.
5. Thayer R E. An improved method for detecting foreign DNA in plasmids of Escherichia Coli. Anal Biochem 1979; 98: 60–63.
6. Suggs S V, Hirose T, Miyake T, Kawashima E H, Johnson M J, Itakura K, Wallace R B. Use of oligodeoxynucleotides for the isolation of specific cloned DNA sequences. ICN-UCLA Symp Dev Biol 1981; 23: 683–693.
7. Wood W I, Gitschier J, Lasky L A, Lawn R M. Base composition-independent hybridization in tetramethylammonium chloride: a method for oligonucleotide screening of highly complex gene libraries. Proc Natl Acad Sci USA 1985; 82: 1585–1588.
8. Crabtree G R, Kant J A. Coordinate accumulation of the mRNAs for the alpha, beta and gamma chains of rat fibrinogen following defibrination. J. Biol Chem 1982; 257: 7277–7279.
9. Crabtree G R, Kant J A. Molecular cloning of cDNAs for a family of coordinately regulated genes: the alpha, beta and gamma chains of rat fibrinogen. J Biol Chem 1981; 256: 9718–9723.
10. Hedrick S M, Cohen D I, Nielsen E A, Davis M M. Isolation of cDNA clones encoding T cells-specific membrane-associated proteins. Nature 1984; 308: 149–153.
11. Yanagi Y, Yoshikai Y, Leggett K, Clark S P, Aleksander I, Mak T W. A human T cell-specific cDNA clone encodes a protein having extensive homology to immunoglobulin chains. Nature 1984; 308: 145–149.
12. Korman A J, Knudsen P J, Kaufman J F, Strominger J L. cDNA clones for the heavy chain of HLA-DR antigens obtained after immunopurification of polysomes by monoclonal antibody. Proc Natl Acad Sci USA 1982; 79: 1844–1848.

7
Molecular cloning of genes: use of antibody screening methods

R. W. Mercer

INTRODUCTION

The identification and isolation of genes from recombinant DNA libraries is often dependent upon the availability of a suitable probe. If part or all of the amino acid sequence of a protein is known, then it is possible to synthesize a specific oligonucleotide for use as a hybridization probe. Unfortunately, often this method is precluded because little or no amino acid sequence is known. However, if antibodies to the polypeptide of interest are available, then the isolation of specific genes from recombinant DNA libraries can often be achieved using the antibodies to detect a polypeptide encoded by the recombinant DNA. Using the λgt11 expression vector originally developed by Young and Davis,[1,2] antibody probes have been successfully used to isolate several genes encoding the immunoreactive polypeptides.[3-8]

The λgt11 expression vector system is outlined in Figure 7.1. cDNA molecules are inserted into a unique Eco RI cleavage site within the β-galactosidase gene (*lacZ*), 53 base pairs upstream from the translation termination codon. If the cDNA is inserted in the proper orientation and reading frame, the resulting recombinant DNA construct will code for a hybrid protein consisting of all but the 16 carboxy terminal amino acids of β-galactosidase fused to the cDNA polypeptide. To identify individual phage clones containing the hybrid protein of interest, phage plaques on a lawn of *E. coli* cells are overlayed with a nitrocellulose filter. Proteins released by the lysis of cells in the plaque bind to the nitrocellulose where they can be probed with antibody. Antibody binding to hybrid proteins is detected using radiolabeled *Staphylococcus aureus* protein A or a secondary antibody. After autoradiography the individual plaque producing the immunoreactive fusion protein can be identified and isolated. Figure 7.2 shows the results of a screening of a λgt11 library of rat brain cDNA with an antibody to the sodium and potasssium-dependent adenosine triphosphatase (Na,K-ATPase). Figure 7.2A shows the autoradiograph of a filter containing ~5×10^4 plaques, one of which is reactive to the Na,K-ATPase antibody. Figure 7.2B shows an autoradiograph of a filter containing the same phage clone after it has been purified to homogeneity.

The λgt11 expression vector has been designed to overcome several potential problems with the expression of foreign proteins in *E. coli* cells. Generally, the

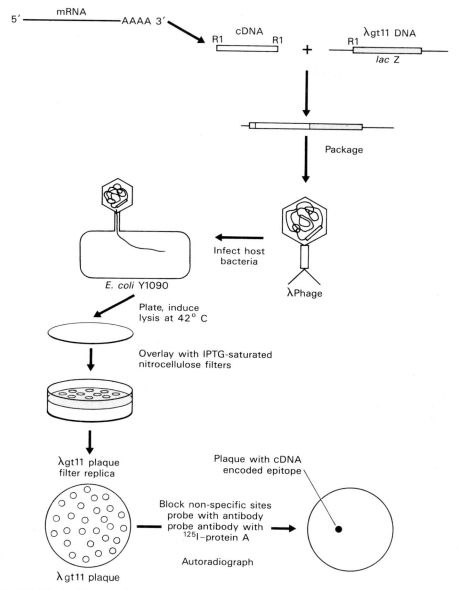

Fig. 7.1 The λgt11 expression system.

level of expression of the fusion protein by the bacterial cell is influenced by the cytotoxicity and stability of the fusion protein. To regulate the expression of potentially harmful hybrid proteins, the *E. coli* strain Y1090 is used. This strain contains multiple copies of the *lac* repressor gene, *lac* I, and therefore, large amounts of *lac* operon repressor, which minimizes the *lac* Z-directed expression

102   MOLECULAR GENETICS

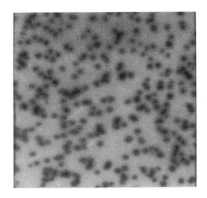

A. Primary Screen   B. Plaque Purified

Fig. 7.2 Immunoscreening of a λgt11 cDNA library. Panels A and B are photographs of autoradiographs of two filters from a typical antibody screening of a λgt11 cDNA library. Panel A shows a single immunoreactive plaque from a plate containing approximately 5×10^{-4} plaques. Panel B shows the autoradiograph of a filter containing the same phage clone after it has been purified to homogeneity.

of fusion protein during initial plaque formation. Once sufficient cells are infected, expression of fusion protein is induced by inactivating the *lac* repressor with isopropyl β-D-thiogalactopyranoside (IPTG). Although the production of the fusion protein can be efficiently induced, proteases in the bacterial cell will often degrade the abnormal protein. To increase the stability of the fusion protein in the bacterial cell, Y1090 is also defective in the production of the *lon* protease. In Y1090 *E. coli* the production of the β-galactosidase cDNA hybrid protein, induced by the *lac* Z promoter and stabilized by the absence of the *lon* protease, can account for a significant percentage of the intracellular protein.

The details of the synthesis of cDNA and the construction of λgt11 libraries are described in Chapter 5 of this volume. Currently λgt11 cDNA libraries from several sources are also commercially available. Several points should be considered in the decision to screen a particular cDNA library. If the cDNA is prepared from a different source than the antigen to which the antibody is directed, their compatibility should be assessed. Usually antibodies that are immunoreactive with antigens on protein blots will produce adequate signals in the λgt11 screening procedure. Also, the likelihood for success in screening a cDNA library is increased if the cDNA is made from a tissue which expresses large amounts of the antigen.

As with other types of library screening, the quality of the probe is also extremely important. An antibody that reacts to several different polypeptides will complicate the analysis of immunoreactive clones. Also, polyclonal antibodies, since they usually recognize several epitopes, tend to be better probes than monoclonal antibodies. It is also essential that the antibody recognize core polypeptide. The bacteria do not have the ability to post-translationally modify the cDNA directed polypeptides, therefore, antiserum against carbohydrate moieties will not detect phage clones expressing the unmodified polypeptide.

ANTIBODY SCREENING OF λgt11 cDNA LIBRARY

Materials:
λgt11 library
E. coli Y1090 [ΔlacU169 proA⁺Δ lonaraD139 strA supF[trpC22::Tn10] (pMC9)]
NZY broth: 10 g/l NZ Amine (Sheffield); 5 g/l Bacto-Yeast Extract (Difco); 5 g/l NaCl; 2 g/l $MgCl_2.6H_2O$; 1.5 ml/l 3 N NaOH
NZY Agar: NZY broth + 15 g/l Bacto-Agar (Difco) + 50 mg/l ampicillin (Sigma)
NZY top agarose: NZY broth + 7.5 g/l agarose (Sigma)
TBS: 8.76 g/l NaCl; 50 mM Tris-HCl, pH 8.0
SM buffer: 5.8 g/l NaCl; 2 g $MgSO_4.7H_2O$; 50 mM Tris-Cl, pH 7.5
sopropyl β- D-thiogalactopyranoside (IPTG) (Sigma): 10 mM, store at −20°C
Bovine serum albumin (BSA) (Sigma)
^{125}I-labeled *Staphylococcus aureus* protein A, specific activity 30 mCi/mg (Amersham)
nitrocellulose filter circles (Millipore or Schleicher and Schuell), 0.45 μm pore size, 150 mm
Triton X-100
Antibody or serum (anti-*E. coli* absorbed)

Plating

1. Place 5×10^4 plaque forming units (in ≤ 50 μl SM buffer) into a sterile test tube. Add 0.2 ml fresh overnight culture of *E. coli* Y1090 grown in NZY broth containing 0.2% maltose and 50 mg/l ampicillin. Incubate for 15 minutes at 37°C.
2. Add 7 ml of molten (50°C) top agarose, pour onto a *dry*, 150 mm NZY-ampicillin plate and gently spread over surface. Invert and incubate for 5–6 hours at 42°C.

After 6 hours plaques should have a diameter of 1–1.5 mm and be just starting to make contact with one another. The plate should not show confluent lysis. Generally, enough plaques should be screened so that each recombinant plaque is represented 2–3 times.

Screening

1. Saturate nitrocellulose filters in 10 mM IPTG. Remove filters and allow to air dry. Number the dry filters with a soft pencil.
2. Overlay plates with nitrocellulose filters, being careful not to trap air bubbles. Incubate for 2 hours at 37°C. Mark the filter in three or more asymmetric locations by stabbing through it and into the agar beneath it with a 18 gauge needle that has been dipped in waterproof black drawing ink.

3. After 30–60 seconds carefully remove the filter and immerse it, plaque side up, in a dish containing TBS. Shake off any top agarose attached to filter. Incubate at room temperature for 10 minutes. Repeat with duplicate filter as described in Step 2.
4. Incubate filters in TBS + 3% BSA (10 ml/filter) overnight at 4°C.
5. Pre-absorb antibody with bacterial lysates prepared from host bacteria. To prepare lysate, recover bacteria from one liter of confluent culture. Resuspend bacteria in 10 ml of distilled water in a 50 ml plastic conical tube. Place suspension in boiling water bath for 5–10 minutes. Use 1 ml of bacterial lysate to absorb 100 ml of diluted antibody. Absorb for 2 hours at 4°C. After pre-absorption remove the bacterial debris by centrifugation at 2000 g for 30 minutes. Store unused lysate at −20°C.
6. Incubate filters in TBS + 3% BSA + 1:100 dilution of serum or IgG (IgG initially at 10 mg/ml). Incubate for 2 hours at room temperature. Rotate filters every 30 minutes. The antibody solution can be reused several times. To prevent bacterial growth in the solution add sodium azide to a concentration of 0.1%; store at 4°C.
7. Wash: 10 minutes in TBS (25 ml/filter) at room temperature
   10 minutes in TBS + 0.1% Triton X-100
   10 minutes in TBS
8. Incubate filters in TBS + [^{125}I] protein A (approximately 10^6 cpm/filter) for 1 hour at room temperature. Rotate filters every 15 minutes.
9. Set up three plastic boxes or glass trays containing 25 ml/filter TBS. Immerse filter successively in trays, collecting filters in last tray.
   Wash: 10 minutes in TBS (25 ml/filter) at room temperature
   10 minutes in TBS + 0.1% Triton X-100
   10 minutes in TBS
10. Dry filters on paper at room temperature. Tape the filters, plaque sides up, onto sheets of Whatman 3MM paper. Arrange duplicate filters next to each other. To align the autoradiograph to the filters, mark the 3MM paper in several locations with radioactive ink or with an autoradiography marker (UltEmit, New England Nuclear). Cover the 3MM paper and filters with Saran wrap and expose to Kodak XAR-5 X-ray film overnight at −70°C using Cronex intensifying screens (Dupont).

Positive signals that are on both filters are isolated by removing an agar plug from the corresponding region of the plate with the wide end of a Pasteur pipet. Incubate the agar plug in 1 ml of SM buffer with 2 drops of chloroform for several hours. The titer of this solution should be approximately $1-2 \times 10^7$ plaque forming units/ml. Plate 500–1000 plaques on a 90 mm NZY-ampicillin plate and repeat the screening procedure until a single immunoreactive plaque can be isolated. Often, as a positive control, it is useful to include in the screening procedure a nitrocellulose filter on which dilutions of the antigen have been spotted.

Once a single positive plaque has been isolated a high titer phage stock should be made. To amplify the immunoreactive phage the bacterial strain Y1088 is used. Y1088, in addition to producing the *lac* repressor, is also $hsdR^-hsdM^+$, rendering it defective in host directed modification of foreign DNA.

PREPARATION OF PLATE LYSATE STOCK

Materials:
  E. coli Y1088 [*supE supF metB trpR hsdR$^-$hsdM$^+$ tonA21 strA ΔlacU169 proC::Tn5*] (pMC9)
  NZY-ampicillin plates
  NZY top agarose
  SM buffer
  chloroform

Protocol
1. Plate 5×10^4 plaque forming units (in ≤50 μl SM buffer) onto a *dry*, 150 mm NZY-ampicillin plate as described above using the host bacterium Y1088. Incubate for 8–10 hours at 42°C.
2. After confluent lysis, add 10 ml of SM buffer. Incubate at room temperature for several hours with slow, gentle shaking.
3. With a pipet remove as much of the SM buffer as possible and place into a glass or polypropylene test tube. Rinse the plate with 2 ml of fresh SM buffer and combine it with the first fraction.
4. Centrifuge at 2000 g for 5 minutes to remove bacterial debris.
5. Remove the supernatant to a fresh tube and add 0.1 ml of chloroform. Seal the tube with parafilm and store at 4°C. The titer of the plate stock is generally $10^{10}-10^{11}$ pfu/ml.

CHARACTERIZATION OF β-GALACTOSIDASE cDNA FUSION PROTEIN

After the positive phage clone has been purified it is useful to verify the immunoreactivity of the fusion protein. Preparative quantities of the β-galactosidase cDNA hybrid protein can also be isolated to be used as an antigen or to affinity purify antibodies. Relatively large quantities of the hybrid protein can be prepared by expressing the recombinant as a lysogen. To produce lysogens the bacterial strain Y1089 is used. In addition to producing the *lac* repressor and being deficient in the *lon* protease, Y1090 also has a mutation (*hflA150*) that increases the frequency in which the phage enter the lysogenic growth cycle. Thus, fusion protein can be prepared by growing the lysogen to high density followed by the induction of the β-galactosidase cDNA hybrid protein with IPTG.

PREPARATION OF λgt11 LYSOGENS IN Y1089

Materials:
E. coli Y1089 [ΔlacU169 $proA^+\Delta lon$ araD139 $StrAhflA$150[chr::Tn10] (pMC9)]
NZY-ampicillin plates
NZY broth
10 mM MgSO$_4$
TNE buffer: 10 mM Tris-Cl, pH 8.0, 100 mM NaCl, 1 mM EDTA
0.5 M IPTG

Protocol
1. Grow Y1089 in NZY broth containing 50 μg/ml ampicillin and 0.2% maltose overnight. A 1:10 dilution of the overnight culture should have an OD$_{600}$ of between 0.7–1.2.
2. Centrifuge overnight suspension at 2000 g for 10 minutes. Resuspend pellet in 10 mM MgSO$_4$ to a final concentration of 2.5 OD$_{600}$/ml. Assume 1 OD$_{600}$ = 8 × 10^8 cells/ml.
3. Dilute bacterial suspension 1:100 to a final concentration of 2 × 10^7 cells/ml. Place 1 × 10^7 and 2 × 10^7 pfu into test tubes; add 0.1 ml of the bacterial suspension. Incubate for 15 minutes at 30°C.
4. Using NZY broth, make 10^{-3}, 10^{-4}, dilutions of the phage-cell mixture. Plate 50 and 100 μl of each onto a NZY-amp plate. Spread plates and incubate right side up at room temperature for 15 minutes.
5. Invert plate and incubate overnight at 30°C.
6. Using sterile toothpicks, pick 70–90 colonies onto 2 NZY-amp plates. Incubate one plate at 30°C and the other at 42°C for 12 hours. Lysogens should grow at 30°C but not at 42°C.
7. Streak out lysogens to obtain single colonies; grow at 30°C.

PREPARATION OF CELL FREE LYSATES FROM LYSOGENS

1. Grow 10 ml of overnight cultures of each lysogen at 30°C for 12 hours.
2. To verify the stability of the lysogen inoculate 2 NZY-amp plates with 25 μl of the culture medium. Incubate one plate at 30°C and the other at 42°C. Use cultures which grow only at 30°C.
3. Inoculate 2 150 ml NZY broth cultures in a 500 ml flask with 5.0 ml of overnight culture. Grow at 30°C until OD$_{600}$ = 0.5.
4. Rapidly shift temperature to 42°C; incubate for 30 minutes.
5. To one culture of each lysogen add 1.5 ml of filter sterilized 0.5 M IPTG (final concentration = 5 mM). Incubate at 38°C for 1.5–2 hours.
6. Centrifuge suspension at 2000 g at room temperature for 10 minutes. Carefully resuspend pellet in 5.0 ml of ice-cold TNE buffer.
7. Rapidly freeze the mixture in a dry ice-ethanol bath. Thaw the suspension and shear the DNA by extruding the suspension 5 times through a 19 gauge needle.

GENE CLONING WITH ANTIBODIES   107

8. Determine protein concentration of suspension. Store lysates frozen at −20°C.

Protein from IPTG-induced and uninduced bacterial lysates can be compared by SDS polyacrylamide gel electrophoresis. Since the fusion protein is a hybrid, consisting of bacterial β-galactosidase and the cDNA directed polypeptide, it should be immunoreactive both with the antiserum used to identify it and an antiserum to β-galactosidase. Figure 7.3 shows bacterial lysates from a lysogen containing a Na,K-ATPase β-subunit cDNA clone. The Coomassie blue staining pattern of the lysates is shown in Figure 7.3A. The addition of IPTG to the wild-type λgt11 lysogen induces the expression of the 116 kd β-galactosidase (identified by the lower arrowhead in panel A). The addition of IPTG to the λNKβ lysogen, however, induces the expression of a 145 kd fusion polypeptide (upper

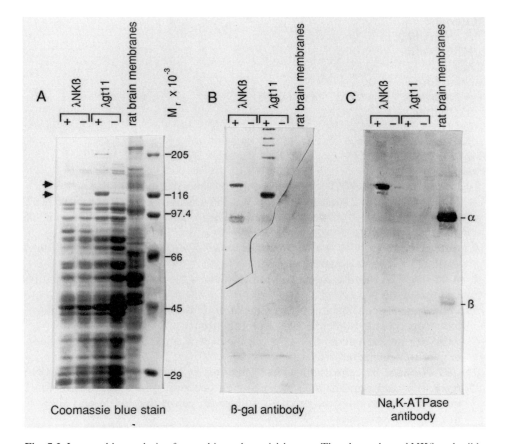

Fig. 7.3 Immunoblot analysis of recombinant bacterial lysates. The phage clones λNKβ and wild-type λgt11 were used to lysogenize *E. coli* Y1089. Production of fusion protein was induced (+) or not induced (−) with isopropyl β-D-thiogalactopyranoside. Lysogen proteins were separated by SDS polyacrylamide gel electrophoresis and transferred to nitrocellulose. (A) Coomassie blue staining of lysogen proteins. (B) Reactivity with β-galactosidase antibody followed by treatment with alkaline phosphatase conjugated to protein A. (C) Reactivity with Na,K-ATPase antibody followed by treatment as in panel B.

arrowhead). As shown in Figures 7.3 (B and C), the 145 kd polypeptide is immunoreactive to both the β-galactosidase and the Na,K-ATPase antibodies, while native β-galactosidase is immunoreactive to only the β-galactosidase antibody. These results demonstrate that the Na,K-ATPase antibody reacts with only the cDNA directed fusion protein and not with bacterial or phage polypeptides.

To isolate the cDNA coding for the identified epitope, DNA from the recombinant λgt11 must be prepared. This is accomplished by isolating bacteriophage from a liquid culture of infected Y1088 *E. coli*.

PREPARATION OF λ BACTERIOPHAGE

Materials:
  E. coli Y1088
  NZY broth
  NaCl
  chloroform
  polyethylene glycol MW 8000 (Sigma)
  SM buffer
  CsCl

Protocol
1. Prepare 1 liter of NZY broth in a 4 liter flask. Warm to 42°C.
2. To 10^7 pfu add 0.2–0.3 ml of fresh Y1088 overnight culture grown in NZY broth containing 0.2% maltose and 50 mg/l ampicillin. Incubate for 15 minutes at 37°C.
3. Add bacteria to NZY broth, incubate at 42°C with shaking for 8–12 hours until cell debris is apparent.
4. Add 30 g NaCl and 2 ml of chloroform, shake for an additional 15 minutes at 42°C.
5. Centrifuge suspension for 10 minutes at 10 000 rpm at 4°C in a GSA rotor. Save supernatant by decanting through cotton gauze.
6. Add 70 g polyethylene glycol, precipitate on wet ice for at least one hour, or place in refrigerator overnight.
7. Collect phage by centrifuging for 30 minutes at 11 000 rpm at 4°C in a GSA rotor. Discard supernatant. Repeat with one dry spin at 5000 rpm for 5 minutes. Remove any supernatant and invert centrifuge bottle and allow to drain for 15 minutes.
8. Resuspend pellet in 23 ml of SM buffer.
9. To phage suspension add 16.79 g desiccated CsCl.
10. Load into two 16 × 76 mm polyallomer Quick-seal centrifuge tubes, centrifuge at 30 000 rpm for 20 hours at 15°C in a Ti50 rotor.
11. Remove the white phage band by puncturing the side of the centrifuge tube with a 18 gauge needle attached to a 5 ml syringe. Placing the tube against

a dark background and shining a light from above often aids in visualizing the band.

EXTRACTION OF DNA FROM λ BACTERIOPHAGE

Materials:
100 × STE buffer: 2 M Tris-Cl, pH 7.5, 1 M NaCl, 100 mM EDTA
Adjust pH with concentrated HCl.
5 M NaCl
1 M Tris-Cl, pH 8.0
1 M $MgCl_2$
0.5 M EDTA
10 mg/ml proteinase K
10% SDS
TE saturated phenol
chloroform
3 M Na acetate

Protocol
1. For each ml of phage suspension from the CsCl gradient add:
   2 µl 5 M NaCl
   50 µl 1 M Tris-Cl, pH 8.0
   10 µl 1 M $MgCl_2$
   40 µl 0.5 M EDTA
   5 µl 10 mg/ml proteinase K
   50 µl 10% SDS
2. Dialyze with shaking against 4 liters of 1 × STE buffer for 1 hour at 37°C.
3. Add an equal volume of TE equilibrated phenol. Gently vortex and centrifuge at 2000 g for 5 minutes. Remove upper aqueous layer to a fresh tube. Repeat phenol extraction.
4. Extract the aqueous phase as above with a 50:50 mixture of phenol and chloroform.
5. Extract the aqueous phase as in Step 3 with an equal volume of chloroform. Add $\frac{1}{10}$ volume of 3 M Na acetate. Add 2 volumes of 100% ethanol, mix, and allow to stand for several hours at −20°C. The DNA should be visible at this point.
6. Centrifuge at 2000 g for 5 minutes. Wash pellet in 70% ethanol, followed by a wash in 100% ethanol. Dry the pellet briefly in a vacuum desiccator. Resuspend pellet in 1–2 ml of TE buffer. Determine concentration of DNA spectrophotometrically.

Once the phage DNA has been purified the cDNA insert can be isolated by Eco RI endonuclease digestion (see Ch. 2). The cDNA can then be used as a probe in RNA and DNA filter hybridization or to identify related or overlapping cDNA clones in the library. The cDNA insert can also be used for nucleotide sequencing, or to hybrid select mRNA.

REFERENCES

1 Young R A, Davis R W. Efficient isolation of genes by using antibody probes. Proc Natl Acad Sci USA 1983; 80: 1194–1198.
2 Young R A, Davis R W. Yeast RNA polymerase II genes: isolation with antibody probes. Science 1983; 222: 778–782.
3 Schwarzbauer J E, Tamkun J W, Lemischka A R, Hynes R O. Three different fibronectin mRNAs arise by alternative splicing within the coding region. Cell 1983; 35: 421–431.
4 Landau N R, St John T P, Weisman I L, Wolf S C, Silverstone A E, Baltimore D. Cloning of terminal transferase cDNA by antibody screening. Proc Natl Acad Sci USA 1984; 81: 5836–5840.
5 Kopito R R, Lodish H F. Primary structure and transmembrane orientation of the murine anion exchange protein. Nature 1985; 316: 234–238.
6 Mueckler M M, Caruso M C, Baldwin S A et al. Sequence and structure of a human glucose transporter. Science 1985; 229: 941–945
7 Schneider J W, Mercer R W, Caplan M, Emanuel J R, Sweadner K J, Benz E J Jr, Levenson R. Molecular cloning of rat brain Na,K-ATPase α-subunit cDNA. Proc Natl Acad Sci USA 1985, 82: 6357–6361
8 Mercer R W, Schneider J W, Savitz A, Emanuel J, Benz E J Jr, Levenson R. Rat-brain Na,K-ATPase β-chain gene: primary structure, tissue-specific expression, and amplification in ouabain-resistant HeLa C^+ cells. Mol Cell Biol 1986; 6: 3884–3890.

8
Analysis of DNA methylation and DNase I sensitivity in gene-specific regions of chromatin

G. Ginder

INTRODUCTION

The complex mechanisms which control eukaryotic gene expression are only beginning to be understood. The successful molecular cloning and sequencing of a large number of eukaryotic genes has allowed the basic transcription unit to be defined, but the critical regulatory interactions between DNA and the wide variety of nuclear proteins, which together constitute chromatin, remain to be fully elucidated. Two identifiable features that have been found to be closely associated with the nuclear chromatin surrounding actively transcribed genes in hematopoietic cells, as well as other tissues, are hypomethylation of DNA[1] and increased accessibility of the DNA to pancreatic DNase I.[2]

Methylation at the 5-position of cytidine in the dinucleotide CpG is the only well-documented post-synthetic modification of DNA in higher eukaryotes. The precise role that DNA methylation plays in gene regulation remains to be firmly established (for review, see[3]). However, a general inverse correlation between DNA methylation and transcription has been demonstrated for many genes.[1] Among the first and the most thoroughly studied gene systems in which DNA hypomethylation was shown to correlate with expression is the globin gene family.[4-6] Detailed mapping of CpG methylation in the vicinity of the globin genes has shown the tightest inverse correlation between methylation and expression to be located in the 5' upstream regions of a given transcription unit where DNase I hypersensitivity sites and corresponding trans-acting factor binding sites are also located.[7,8] Moreover, *in vitro* methylation of the 5' regions of the cloned human γ-globin gene results in complete inhibition of transcription when the gene is subsequently transfected into cultured cells.[9]

DNase I sensitivity can be divided into two types: generalized (or intermediate level) DNase I sensitivity and DNase I hypersensitivity. Generalized DNase I sensitivity, first described by Weintraub & Groudine,[2] is thought to reflect a basic opening of the chromatin surrounding genes that are either actively transcribed or transcriptionally inactive genes that are either 'poised' for expression or adjacent to actively transcribed sequences (e.g. embryonic globin genes in adult erythroid cells). So-called DNase I hypersensitive sites, on the other hand, are generally found proximal to actively transcribed genes.[10-12] DNase I hypersensitive sites are much more circumscribed and appear to reflect short stretches of DNA around which the usual chromatin structure has been completely disrupted by the binding of soluble proteins, which recent evidence suggests may be tissue-specific trans-acting stimulators of transcription.[7,8] Although most of the originally described DNase I hypersensitivity sites were located at the 5' regions of their respective genes,[12] there are now well-documented hypersensitive sites within and at the 3' ends of specific genes as well.[13] The increasing association between DNase I hypersensitive sites and either cis-acting gene enhancer sequences or trans-acting binding proteins has made the DNase I hypersensitivity assay a widely applicable method for localizing potentially important control regions within the DNA sequences encompassing specific genes.[13]

Thus, taken in concert, assays for DNA methylation and DNase I hypersensitivity can provide important information regarding potential gene control regions for any gene which has been cloned and structurally characterized. Although the specific methodology for DNA methylation assays and DNase I sensitivity assays differ, both are based primarily on digestion of genomic DNA by specific endonucleases followed by separation of DNA fragments by gel electrophoresis and transfer of DNA to a solid phase support membrane for hybridization to labeled DNA containing all or part of the gene sequence under study, i.e. the Southern blot,[14] which is described in detail in Chapter 2.

METHODS

Determination of DNA methylation in the vicinity of specific genes

The use of bacterial restriction endonucleases that cleave at recognition sequences that include the dinucleotide CpG and that are inhibited by the presence of a methyl group at position 5 of the cytosine ring, allows the assessment of methylation in any region of DNA for which a unique sequence probe is available. The typical method involves the use of a pair of restriction enzyme isoschizomers which cleave at the sequence CCGG. The first of these, designated Hpa II (or Hap II), will not cleave the sequence $CC^{me}GG$ while the second, Msp I, will cleave regardless of the CpG methyl group except in rare circumstances.[15-17] While a variety of other restriction enzymes that are inhibited by cytosine methylation are available (see Table 8.1), Hpa II/Msp I, and Sma I/Xma I, which recognize a subset of Hpa II/Msp I sites, are the only pairs that allow direct verification that a specific set of CpG sites are methylated in genomic DNA.

The basic method for mapping methylation sites within and around a specific gene is described below and represented schematically in Figure 8.1.

1. Isolation of DNA

High molecular weight DNA (> 30 kb mean fragment length) can be isolated from cells by a variety of techniques (of Chs. 2, 13).[19] For most tissue culture cells and bone marrow cells when adequate numbers of cells (> 10^8) are available, best results are obtained by preparing nuclei followed by proteinase diges-

Table 8.1 5-Methylcytosine sensitive restriction endonucleases. Derived from McClelland & Nelson.[18]

Enzyme§	Sequence specificity*
Ava I	CPy†MCGPu‡G
Bsu M	CT^{m}CGAG
Hha I (Cfo I)	G^{m}CGG
Msp I	^{m}CCGG
Hpa II	C^{m}CGG
Nae I	GCCGG^{m}C
Pvu I (Xor II)	CGAT^{m}CG
Sma I	CC^{m}CGGG
Tha I	CG^{m}CG/^{m}CGCG
Xma I	C^{m}CCGGG

* The superscript '^{m}' immediately preceding a 'C' indicates methylation of that cytosine base. These enzymes are blocked from restriction when the designated cytosine in the recognition sequence is methylated.
† Py — this base may be either cytidine or thymine.
‡ Pu — this base may be either guanine or adenine.
§ Enzymes in parenthesis share the same recognition site and methyl sensitivity as the enzyme preceding them.

tion and phenol extraction. A detailed description of a reliable method for isolating intact nuclei is given below in Section 2 on DNase I sensitivity. When tissues are in low abundance (10^6 cells), whole cell DNA extraction provides higher yields, though the resultant DNA may subsequently be more difficult to digest with restriction enzymes. Methods for whole cell DNA extraction are described in Chapter 2.

2. Primary restriction digestion

It is generally felt, and it has been our experience, that high molecular weight DNA must be cleaved by one or more 'infrequent cutter' restriction enzymes (i.e. restriction enzymes which require a recognition sequence of 6 or more bases to cut DNA) that are not sensitive to cytosine methylation in order to ensure subsequent complete cleavage of all unmethylated recognition sites by restriction enzymes that are inhibited by 5 methyl cytosine. This is particularly important when one is using enzymes other than Hpa II/Msp I, since in those instances there is no direct 'internal control' for complete digestion by the methyl-sensitive enzyme. A typical reaction would include the following:

50 µg high molecular weight genomic DNA
40 µl 10 × Hind III restriction buffer
100 units Hind III restriction enzyme
Distilled/deionized H_2O QS
400 µl total reaction volume

Incubate at 37°C for 2 hours

In order to monitor for complete restriction digestion, we include a control reaction as follows:

a. 20 μl (or about $\frac{1}{20}$ volume) of restriction digest reaction mixture is removed after 2 hours of incubation.
b. 1 μg of λCI857 DNA (or other suitable plasmid or bacteriophage DNA) is added and the control reaction is vortexed.
c. An additional 100 units Hind III is added to the main reaction mixture.
d. The control and main reactions are co-incubated for an additional 2 hours at 37°C.
e. The control reaction(s) are electrophoresed in a 1% agarose gel with ethidium bromide staining and examined under UV illumination.
f. If the λ marker phage DNA is digested to completion, DNA is isolated from the main reaction as follows:
g. $\frac{1}{40}$ volume of 0.4 M EDTA (final concentration 0.01 M) plus $\frac{1}{50}$ volume of a 5 mg/ml solution of proteinase K (final concentration 100 μg/ml) and $\frac{1}{100}$ volume of 20% N-Lauroyl Sarcosine (final concentration 0.2%) are added and the mixture is incubated for 2 hours at 37°C.
h. The reaction mixture is extracted × 3 with a mixture of neutral phenol, chloroform and isoamyl alcohol (24 : 24 : 1) followed by extraction × 3 with chloroform/isoamyl alcohol (24 : 1).
i. $\frac{1}{10}$ volume of 3 M sodium acetate pH 5.5 is added and the DNA is precipitated with 2 volumes 95% ETOH at −20°C overnight.

3. Restriction with methyl-sensitive enzymes

The second digestion with methyl-sensitive restriction enzyme is carried out essentially in the same manner as the primary digestion. Usually 15–20 μg of each DNA sample provides a sufficient amount for one or two Southern blots. A 5–10-fold excess of the respective restriction enzyme should be added to ensure that there is complete digestion of unmethylated sites. It is also critical to include a control reaction with a suitable phage or plasmid DNA to ensure complete digestion. This is particularly important when using restriction enzymes other than Hpa II/Msp I or Sma I/Xma I since there is no internal control for complete cleavage at all non-methylated sites in cellular DNA.

4. Gel electrophoresis and blotting (see also Ch. 2)

Standard agarose gel electrophoresis can be carried out with gels composed of from 0.8% to 2% agarose in either 1 × TAE (0.04 M Tris acetate, 0.002 M EDTA) buffer or 10 mM sodium phosphate pH 7.0 buffer. Sodium phosphate buffer offers the advantage of allowing rapid separation of higher molecular weight DNA fragments (5–20 kb), but suffers a slight disadvantage in that an overloading effect, resulting in band smearing, occurs at lower DNA amounts than with TAE buffer. It is helpful to include one lane of ^{32}P end-labeled phage or plasmid restriction fragment size markers to facilitate the sizing of genomic DNA restriction fragment signals following blotting and autoradiography.

Transfer of the gel to a membrane or filter for subsequent hybridization may be carried out by a variety of techniques including the original Southern method[14] or a number of variations. We currently use the alkaline transfer method of Reed & Mann[20] in which the agarose gel is treated with 0.1 N NaOH and transferred to a nylon membrane without neutralization. This method is fast and simple, and the nylon membrane offers several advantages over nitrocellulose including superior DNA retention and stability during successive manipulation and rehybridization.

5. Probe synthesis and hybridization
Various methods of probe synthesis have been described including nick translation[19] and synthesis of ^{32}P-labeled cRNA.[21] We have found that the most straightforward method for blot hybridizations is to use either standard nick translation[19] or the random primer method of Vogelstein.[22] The latter method allows labeling of very small quantities of specific DNA fragments which can actually be separated by electrophoresis in low melting temperature agarose gels and labeled without isolation or subcloning the individual DNA fragments.[22] Hybridization probes in the range of $2-9 \times 10^8$ cpm/mg are generally required for high quality, low background autoradiograms.

Several filter hybridization techniques have been described,[19] and they involve either aqueous buffers with high temperature incubation or formamide based buffers at lower hybridization temperatures. Our typical hybridization reaction is carried out as follows:

a. Blotted filters are pre-incubated for 60 minutes at 42°C in hybridization buffer without probe in a Seal-a-Meal bag.
Hybridization buffer:
50% deionized formamide
5 × SSC (0.75 M NaCl, 0.075 M sodium citrate)
0.038 M [Na] PO$_4$
5 × Denhardt's solution.[19]
200 µg/ml denatured salmon sperm DNA
b. Buffer is removed and fresh hybridization buffer containing denatured hybridization probe 5×10^6 cpm/ml of buffer and approximately 4 ml of this buffer and 1 ml of 50% dextran sulfate per 100 cm^2 of filter are mixed and added and the bag is resealed.
c. The filter is incubated at 42°C with gentle agitation for 24–36 hours while maintaining the filters in a flat position to ensure uniform exposure to probe.
d. The filter is removed from hybridization mixture, blotted with paper towels and washed for 15 minutes each as follows:
Wash #1 — 2 × SSC, 10 mM Na pyrophosphate, 0.1% SDS at 50°C (2 washes)
Wash #2 — 0.1 × SSC, 0.1% SDS at 50°C (2 washes)
Wash #3 — 0.1 × SSC at 65°C (1 wash)
(The final high stringency wash is optional and should not be used for filter reactions in which hybridization probes are less than 150 base pairs in length.)
e. Filters are finally air dried and exposed to high sensitivity X-ray film (Kodak

6. Interpretation of restriction map analysis of methylation sites

Figure 8.1 schematically illustrates the mapping of methylation of C^mCGG sites in the chick embryonic ρ-globin gene in embryonic erythroid cells in which the gene is transcribed and in adult erythroid cells in which it is not transcribed.[23,24] A primary digestion of genomic DNA with Hind III releases a 4.6 kb fragment containing the ρ gene and about 3 kb of flanking DNA. When the DNA from embryonic erythroid cells is subsequently digested with Hpa II or Msp I, as illustrated in panel B on the left side of Figure 8.1, all five CCGG sites in the Hind III fragment are cleaved by virtue of the absence of any ^mCpG. In contrast, as shown on the right-hand side of Figure 8.1, all five CCGG sites in the adult erythroid cell DNA are methylated so that Hpa II does not cleave at any of these sites; while Msp I, which is not inhibited by C^mCGG, cleaves at all five sites. When the DNAs are subsequently electrophoresed in an agarose gel, the ethidium stained photographs of bulk DNA depicted in Figure 8.1, panel C, show that in both cell types the overall methylation at C^mCGG sites is about equal and considerable, since the mean fragment cleavage size for Hpa II is considerably larger than for Msp I. When the DNA in the gels is transferred to nitrocellulose membranes and hybridized to a ^{32}P-labeled probe specific for the ρ-globin gene as shown in panel E, the gene fragments from Msp I and Hpa II digests are identical in the DNA from embryonic erythroid cells, reflecting the lack of any C^mCGG methylation. In contrast, DNA from the non-expressing adult erythroid cells yields an intact 4.6 kb Hind III fragment when cleaved with Hpa II but yields a complete limit digest identical to embryonic erythroid DNA when cleaved by Msp I. Since all CCGG sites which lie 5' upstream from the ρ hybridization probe are methylated in this example, a single 3' probe allows one to map all of the ρ gene methylations at CCGG sites in the adult erythroid cell DNA. However, to be sure that sites 1 and 2 are truly unmethylated in embryonic erythroid DNA in this example, one would have to rehybridize the Southern blot in panel E with a 5' probe which spans these sites. Thus, in order to map multiple methylation sites in a particular gene region, it is usually necessary to hybridize with several unique sequence probes to allow unequivocal assessment of all sites. When one is mapping methylation sites with methyl-sensitive enzymes such as Hha I (G^mCGC), for which there is no isoschizomer which will cleave at the methylated recognition site, it is necessary to include a restriction digest of cloned DNA corresponding to the genomic DNA region that is being mapped in order to determine the presence of all potential cleavage sites.

7. Mapping CpG methylation sites by genomic sequencing

Since many potentially methylated CpG dinucleotides in a given gene region will not lie within the recognition sequence of any restriction endonuclease that is sensitive to 5 methyl cytosine, it may not be possible to use restriction site mapping to study methylation of the relevant control regions of some genes. The

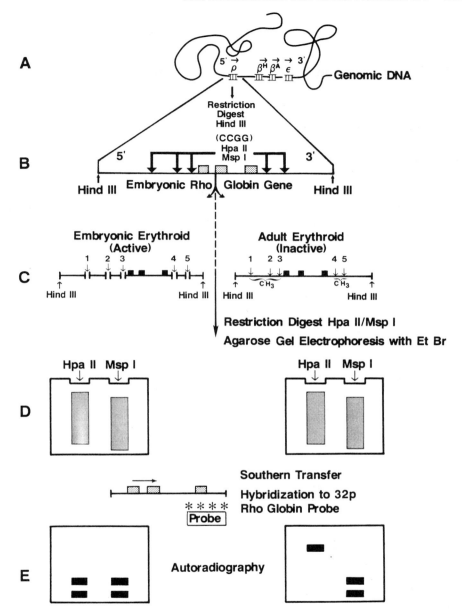

Fig. 8.1 Strategy for mapping cytosine methylation at CpG sites in specific gene regions. In the example depicted, total cellular DNA was extracted from either adult chick erythroid cells in which the rho (ρ) embryonic globin gene is transcriptionally inactive (right side), or 5-day embryonic erythroid cells in which the ρ gene is transcribed (left side), and analyzed for methylation at CCGG sites in the DNA sequences encompassing the ρ gene. Vertical arrows pointing downward (↓) designate Hpa II/Msp I restriction enzyme recognition sites, upward arrows (↑) depict Hind III restriction sites; and horizontal arrows (→) depict the direction of transcription of the ρ gene.

recent development of an elegant technique known as genomic sequencing[25] makes it possible to examine cytosine methylation at all CpG sites of a given DNA region. While this method is more technically demanding than restriction mapping, it is valuable for detailed studies of the mechanistic role of cytosine methylation in gene regulation. The details of genomic sequencing and its application to the determination of cytosine methylation have been described by Gilbert and colleagues,[26] and will not be reiterated here. The general principle of the method is based on chemical sequencing of DNA coupled with the fact that 5 methyl cytosine, in contrast to cytosine, is not cleaved by hydrazine, so that base positions occupied by 5 methyl cytosine appear as 'gaps' in the sequencing ladder.[27] When eukaryotic genes are cloned and amplified in appropriate *E. coli* host-vector systems, all 5 methyl cytosines in the sequence CpG are lost. Thus, by comparing the sequence encompassing a cloned gene fragment with the genomic sequence of the same region, one can potentially localize any and all methylated cytosines by virtue of their absence in the genomic sequence ladder.

DNase I sensitivity mapping of specific regions of chromatin
Mapping of intact chromatin regions in which DNA sequences are in a conformation which renders them accessible to cleavage by exogenous nuclease requires the ability to isolate intact nuclei and a nuclease which can penetrate the nucleus and which has minimal base sequence specificity. Pancreatic DNase I has served as the standard nuclease in this type of assay since the original observation of preferential nuclease sensitivity of active genes,[2] although other nucleases have also been successfully used to demonstrate the altered chromatin structure associated with DNase I hypersensitivity.[7]

1. Isolation of nuclei
This is a critical step in successful nuclease sensitivity mapping and probably accounts for most difficulties encountered in this assay due to the variability in fragility of nuclei from different tissues and cell types. The following method is fairly gentle and has been successfully applied to numerous cultured cell lines as well as fresh peripheral blood and bone marrow cells.
   Buffers:
Nuclear buffer 1 (NB-1)
   0.32 M sucrose
   0.002 M $MgCl_2$
   0.010 M [K] PO_4, pH 6.8
Nuclear buffer 2 (NB-2)
   0.010 M NaCl
   0.010 M [K] PO_4, pH 6.8
Nuclear buffer 3 (NB-3)
   0.32 M sucrose
   0.001 M $MgCl_2$
   0.05% Triton X-100

0.001 M Pipes (piperazine N, N'[2-ethane sulfate]), pH 6.4
0.1 mM PMSF (Phenylmethylsulfonyl fluoride)

RB Buffer
0.1 M NaCl
0.05 M Tris-Cl, pH 8.0
0.003 M $MgCl_2$
0.1 mM PMSF
0.005 M sodium butyrate, pH 7.0

Note: Sodium butyrate and PMSF should be added immediately before the respective buffers are to be used.

2. DNase I digestion procedure

a. Cells are washed in NB-1 at 10^7 cells/ml × 2 and centrifuged at 1000 g for 6 minutes in a swinging bucket rotor.
b. Cells are resuspended in an equal volume of NB-2 and spun at 800 g for 10 minutes.
c. Cells are then resuspended in 1–2 ml of NB-3 per 10^7 cells and dounced with 5 strokes using pestle *A* in a Kontes all glass dounce homogenizer. Nuclei should be examined under a microscope at this step to ensure complete cell lysis. If lysis is not complete, repeat this step. Some cell types may require higher concentrations of Triton X-100, up to 0.5%, in order to effect complete cell lysis.
d. Lysed cells are then spun down at 800 g for 10 minutes.
e. Nuclei are washed free of cell debris by washing twice in RB buffer 1–2 ml/10^7 nuclei and spun down at 800 g. (Examine nuclei under a microscope to ensure removal of cell debris.)
f. For storage — nuclei are resuspended at 50 OD_{260}/ml with the addition of an equal volume of sterile glycerol and stored at $-80°C$ in small aliquots. Nuclei are generally stable under these conditions in excess of 6 months.
g. For immediate use in DNase I titrations — resuspend nuclei in RB + 1 mM $CaCl_2$ at 10 OD_{260} units/ml.
h. A series of dilutions (1:50, 1:100, 1:200, 1:500, 1:1000, 1:2000, 1:5000) in RB buffer + 1 mM $CaCl_2$ are made from a 1 mg/ml stock solution of pancreatic DNase I (Worthington).
i. 300 µl aliquots of nuclei are pipeted into 2 ml eppendorf type tubes.
j. 60 µl of the appropriate DNase I dilution (0, 1:50, 1:100, etc.) is added to each tube and mixed well by vortexing at low speed.
k. The DNase I digestion reactions are then incubated at 37°C for 30 minutes and quenched by adding EDTA to a final concentration of 0.01 M.

Note: Since the activity of different DNase I preparations may vary, the dilution range and incubation times need to be adjusted empirically.

3. Preparation of nuclear DNA

High molecular weight nuclear DNA is prepared as described above in the previous section on methyl site mapping.

4. Restriction digestion

The procedure for restriction digestion and monitoring for complete digestion is the same as that described above for methylation site mapping. The selection of appropriate restriction enzymes requires a detailed restriction map of the region of the DNA contained in the chromatin region being studied. As illustrated in Figure 8.2, the restriction sites serve as fixed reference points to allow precise positioning of the nuclease hypersensitivity site(s) by indirect end-labeling of the resulting restriction fragment of a known length. For optimum accuracy then, it is necessary to do separate secondary restriction digestion with at least two different enzymes which make cuts flanking a given hypersensitive site.

5. Blotting and hybridization

The procedures for blotting and hybridization reactions are also as described for methylation site mapping. The selection of the appropriate hybridization probes is based on the fact that this type of mapping represents a form of indirect end-labeling. That is, given that a restriction fragment that encompasses the putative hypersensitive site has been identified, one selects probes which hybridize to either the 3' or 5' distal end of the restriction fragment as illustrated in Figure 8.2, in which a 3' end hybridization probe has been selected to label a Hind III fragment encompassing the embryonic ρ-globin gene. By independently hybridizing to both 5' and 3' probes, one can confirm the position of any hypersensitive site(s).

6. Interpretation of DNase I sensitivity assay

As depicted schematically in Figure 8.2, two types of information can be obtained from the DNase I sensitivity assay. The level of overall or intermediate DNase I sensitivity for a given gene is determined by the rate at which the discrete restriction fragment signal for that gene fades out with increasing DNase I treatment. The fade out is due to the extreme heterogeneity of fragment sizes generated by the random cutting which occurs at higher concentrations of DNase I. This is illustrated by the small downward arrows in panel A of Figure 8.2. In the illustration shown, the ρ-globin gene Hind III 4.6 kb hybridization band, derived from an erythroid cell which expresses the gene, has completely disappeared at a DNase I concentration at which the corresponding ρ-globin gene fragment signal remains clearly visible in the autoradiogram derived from a non-erythroid tissue. As a general rule, genes in an active or open chromatin domain are about ten-fold more sensitive to DNase I than genes in an inactive or closed chromatin domain; therefore the discrete restriction fragment bands from genes in active domains should fade out at about a 10-fold lower DNase I concentration in the case of nuclei from erythroid tissues.

The other important information which can be derived from the DNase I titration experiment, is the localization of DNase I hypersensitive sites. In the example illustrated in Figure 8.1, the ρ-globin gene in the intact nucleus (panel A) has a single 5' DNase I hypersensitive site represented by the large vertical arrow. When DNA from DNase I treated nuclei is isolated and cleaved with

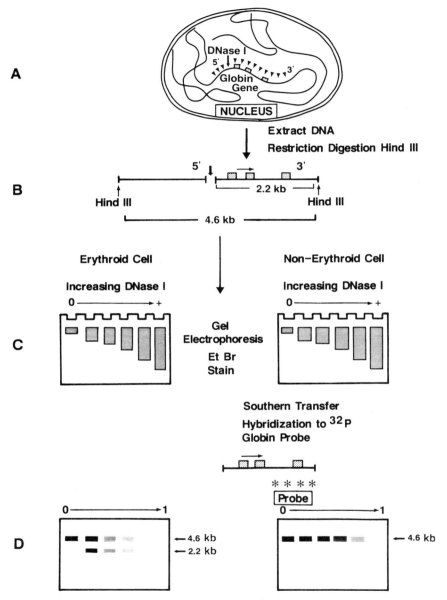

Fig. 8.2 Strategy for determining DNase I sensitivity and DNase I hypersensitive sites in chromatin containing specific genes. In the example shown, nuclei from either erythroid (left-hand side) or non-erythroid (right-hand side) avian cells are isolated and treated with increasing amounts of DNase I. The single DNase I hypersensitive site in this example is designated by a large downward vertical arrow (↓); small downward arrows (▼) indicate random DNase I cleavage sites which are generated at higher concentrations of DNase I; and upward vertical arrows (↑) indicate Hind III restriction sites flanking the ρ-globin gene. The hybridization probe in this example labels the 3′ end of the 4.6 kb Hind III fragment containing the ρ gene under analysis.

Hind III, the ρ gene is contained in a single 4.6 kb fragment. As shown in panel B, the presence of the DNase I hypersensitive site in embryonic erythroid cells results in cleavage of the 4.6 kb Hind III fragment into a 2.2 kb subfragment containing the ρ globin structural gene, while the absence of the hypersensitive site in non-erythroid cell nuclei results in an intact 4.6 kb fragment containing the ρ gene. Thus, after gel electrophoresis, Southern blot transfer, and indirect end-labeling of the Hind III 4.6 kb fragment by hybridization to a ^{32}P-labeled 3′ ρ-globin gene probe, the autoradiograms in panel D are obtained. The autoradiogram on the left, derived from embryonic erythroid cell nuclear DNA, demonstrates both the intact 4.6 kb band and the 2.2 kb sub-band generated by cleavage at the 5′ hypersensitive site; while the autoradiogram on the right, derived from non-erythroid nuclear DNA, shows only the intact 4.6 kb Hind III generated band. Thus, the known position of the 3′ Hind III restriction site serves as the fixed reference point for mapping the position of the 5′ ρ gene hypersensitive site. In order to confirm and more accurately define the position of this site, it would be necessary to repeat this type of analysis with other restriction enzymes that have well-mapped sites in the gene region. Moreover, Figure 8.2 represents a very simple hypothetical case in which a single 5′ DNase I hypersensitive site is present only in cells which actively express the ρ-globin gene. In fact, it has become increasingly apparent that most active gene regions contain multiple nuclease hypersensitive sites, and that some nuclease hypersensitive sites may correlate with the overall chromatin structure of a multigene domain rather than the active transcription of a particular gene.[28] Therefore it is necessary to use multiple 3′ and 5′ probes as well as multiple separate restriction digests to identify and localize all of the potential DNase I hypersensitive sites in a given chromatin domain.

REFERENCES

1 Razin A, Riggs A D. DNA methylation and gene function. Science 1980; 210: 604–610.
2 Weintraub H, Groudine M. Chromosomal subunits in active genes have an altered conformation. Science 1976; 193: 848–856.
3 Razin A, Cedar H, Riggs A D. DNA methylation biochemistry and biologic significance. New York: Springer-Verlag, 1984.
4 McGhee J D, Ginder G D. Site-specific methylation in the vicinity of the chicken β-globin genes. Nature 1979; 280: 419–420.
5 Van der Ploeg L H T, Flavell R A DNA methylation in the human γ, δ, β-globin locus in erythroid and non-erythroid tissues. Cell 1980; 19: 947–958.
6 Shen C K J, Maniatis T. Tissue-specific DNA methylation in a cluster of rabbit β-like globin genes. Proc Natl Acad Sci USA 1980; 77: 6634–6638.
7 McGhee J D, Wood W I, Dolan M, Engel J D, Felsenfeld G. A 200 base pair region of the 5′ end of the chicken adult β-globin gene is accessible to nuclease digestion. Cell 1981; 27: 45–55.
8 Emerson B M, Felsenfeld G. Specific factor conferring nuclease hypersensitivity at the 5′ end of the chicken adult β-globin gene. Proc Natl Acad Sci USA 1984; 81: 95–99.
9 Busslinger M, Hurst J, Flavell R A. DNA methylation and regulation of globin gene expression. Cell 1983; 34: 197–206.
10 Wu C. The 5′ ends of Drosophila heat shock genes in chromatin are hypersensitive to DNase I. Nature 1980; 286: 854–860.

11 Stalder J, Larsen A, Engel J D, Dolan M, Groudine M, Weintraub H. Tissue specific DNA cleavages in the globin chromatin domain introduced by DNase I. Cell 1980; 20: 451–460.
12 Elgin S C R. DNase I-hypersensitive sites of chromatin. Cell 1981; 27: 413–415.
13 Parslow T, Granner D K. Chromatin changes accompany immunoglobulin kappa gene activation: a potential control region within the gene. Nature 1982; 299: 449–451.
14 Southern E M. Detection of specific sequences among DNA fragments separated by gel electrophoresis. J Mol Biol 1975; 98: 503–517.
15 Waalwijk C, Flavell R A. Msp I, an isoschizomer of Hpa II which cleaves both methylated and unmethylated Hpa II sites. Nucleic Acids Res 1978; 5: 3231–3236.
16 Singer J, Roberts-Ems J, Riggs A D. Methylation of mouse liver DNA by means of the restriction enzymes Msp I and Hpa II. Science 1979; 203: 1019–1020.
17 Busslinger M, deBoer E, Wright S, Grosveld F G, Flavell R A. The sequence GGC^{m}CGG is resistant to Msp I cleavage. Nucleic Acids Res 1983; 11: 3559–3569.
18 McClelland M, Nelson M. The effect of site-specific methylation on restriction endonuclease digestion. Nucleic Acids Res 1985; 13(suppl): r201–r207.
19 Maniatis T, Fritsch E F, Sambrook J. Molecular cloning — a laboratory manual. New York: Cold Spring Harbor Laboratory, 1982.
20 Reed K C, Mann D A. Rapid transfer of DNA from agarose gels to nylon membranes. Nucleic Acids Res 1985; 13: 7207–7221.
21 Melton D A, Kreig P A, Rebagliati M R, Maniatis T, Zinn K, Green M R. Efficient in vitro synthesis of biologically active RNA and RNA hybridization probes from plasmids containing a bacteriophage SP6 promoter. Nucleic Acids Res 1984; 12: 7035–7054.
22 Feinberg A P, Vogelstein B. A technique for radiolabeling DNA restriction fragments to high specific activity. Anal Biochem 1983; 132: 6–13.
Addendum, Anal Biochem 1984; 137: 266–267.
23 Groudine M, Peretz M, Weintraub H. Transcriptional regulation of hemoglobin switching in chicken embryos. Mol Cell Biol 1981; 1: 281–288.
24 Ginder G D, Whitters M W, Pohlman J K, Chase R W. In vivo demethylation of chicken embryonic β-type globin genes with 5-azacytidine. In: Stamatoyannopoulas G, Neinhuis A W, (eds). Globin gene expression and hematopoietic differentiation. New York: Alan R Liss, 1983: p 501–510.
25 Church G M, Gilbert W. Genomic sequencing. Proc Natl Acad Sci USA 1984; 81: 1991–1995.
26 Nick H, Bowen B, Ferl R J, Gilbert W. Detection of cytosine methylation in the maize alcohol dehydrogenase gene by genomic sequencing. Nature 1986; 319: 243–246.
27 Ohmori H, Tomizawa J I, Maxam A M. Detection of 5 methylcytosine in DNA sequences. Nucleic Acids Res 1978; 5: 1479–1485.
28 Forrester W C, Thompson C, Elder J T, Groudine M. A developmentally stable chromatin structure in the human β-globin gene cluster. Proc Natl Acad Sci USA 1986; 83: 1359–1363.

9
DNA analysis by M13-dideoxy sequencing
M. Poncz

BACKGROUND

Since first introduced approximately ten years ago, the chain-termination technique using dideoxynucleotides has proven extremely applicable for rapid DNA sequencing. To a large measure, the success of dideoxy sequencing has been based on the development of useful M13 phage constructs and DNA sequencing strategies. However, this technique has a major limitation in that only the sequence within several hundred bases of the universal primer site (see below) can be determined. To sequence the remainder of the insert, numerous random sequencing strategies have been developed. Such approaches are sufficient for short fragments of DNA (less than a kb), but for longer inserts the number of separate reactions needed and the sophistication of data storage and analysis increase rapidly. Therefore, newer non-random sequencing strategies have been developed for analyzing larger inserts. Below, the basis of M13-dideoxy sequencing is presented, followed by a description of several of the strategies we have found useful in the analysis of larger DNA regions.

In dideoxy sequencing, 5' to 3' DNA synthesis is accomplished using a single stranded template which is annealed to a short complementary DNA primer (Fig. 9.1). Until recently, Klenow, the large fragment of *E. coli* DNA polymerase which has 5' to 3' polymerase activity, was the enzyme most often used to extend the primer. Recently, we have begun to use a modified T_7 polymerase as it offers the ability to sequence further with less artifact than Klenow. The use of both enzymes is described below. The extension reaction contains all four deoxynucleotide triphosphates (dNTPs) and one of the four dideoxy substrates (ddNTP) that lack both the 2'- and 3'-hydroxyl groups. ddNTPs can be incorporated into a growing DNA strand, but block further 3' extension of the chain due to a lack of a 3'-hydroxyl group which normally participates in a phosphodiester bond with the α-phosphate group of the next incoming deoxynucleotide triphosphate. In dideoxy sequencing, four separate reactions are run, each of which contains a different ddNTP. If the appropriate ratio of dNTP/ddNTP is used, chain termination will occur randomly at every position within the sequence where that nucleotide is incorporated. To determine the DNA sequence, one of the four dNTPs is radiolabeled (either $\alpha[^{32}P]$ or $[^{35}S]dNTP$). For each reaction mixture, electrophoresis through a thin urea/polyacrylamide denaturing gel followed by autoradiography yields a family of bands, representing the incorporation of the

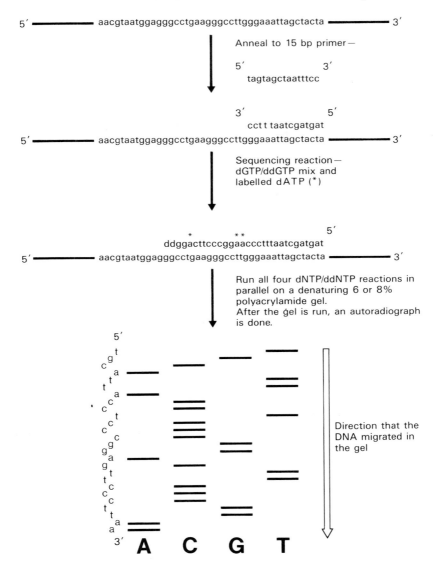

Fig. 9.1 Hypothetical example of dideoxy sequencing using a single stranded template and a 15 bp complementary primer.

particular ddNTP at the various positions in the sequence where that nucleotide occurs. By running all four dNTP/ddNTP reactions in parallel, a ladder of the sequence under study is produced.

Filamentous M13 phage constructs are very useful for providing single stranded templates for dideoxy sequencing. This phage exists as a double stranded form (replicative form, RF) inside *E. coli* and as a single stranded fila-

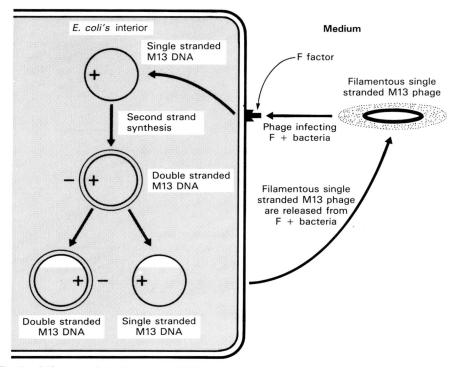

Fig. 9.2 Life cycle of the filamentous M13 phage.

mentous form in the medium (see Fig. 9.2). Only one of the two DNA strands of M13 is excreted into the medium (denoted by convention as the positive (+) strand), providing an easy way to obtain large amounts of pure single stranded DNA (ss DNA). Through genetic engineering, the *lac* promoter and the first 145 amino acid residues of the β-galactosidase gene have been inserted into M13. A short multiple cloning site (MCS) containing the sequences of a variety of unique restriction endonucleases has been inserted into the β-galactosidase gene without interfering with enzymatic function. Depending on the exact construct, the MCS contains different restriction sites in different orientations to the universal primer site. Digestion of RF M13 at one of these sites and insertion of an appropriate fragment of DNA interrupts the expression of the β-galactosidase gene. The filamentous phage has no insert size limitation, although spontaneous deletions in inserts greater than 3–5 kb have been reported. Detection of M13 phage with a DNA insert subcloned into the MCS is accomplished using a chromogenic assay. M13 plaques with the insert remain clear on plates that contain the gratuitous inducer, isopropylthiogalactopyranoside (IPTG), and the chromogenic substrate 5-bromo-4-chloro-3-indolyl-β-D-galactoside (X-GAL) because the interrupted β-galactosidase gene can no longer code for functional enzyme. M13 phage with no insert are detected as blue plaques due to IPTG induction of β-galactosidase activity; the enzyme cleaves the X-GAL to produce blue plaques.

At the present time, we are using M13mp18 and M13mp19 phage constructed by J Messing.[1] These phage have identical MCS in opposite orientations. The advantage of using these two phage is the availability of a great number of unique restriction endonuclease sites in the MCS, allowing significant freedom in the selection of DNA fragments to be inserted. The primer used for most sequencing in our laboratory is the so-called universal primer, a synthetic short oligonucleotide complementary to a region of the β-galactosidase gene immediately downstream to the MCS and which can serve as a primer for all of the M13 constructs. The bacterium presently used is JM107, which is F+. (The F factor is necessary for the filamentous single stranded M13 to enter the cell and is found on an episome in JM107.) To maintain the F factor, the JM107 must be maintained on minimal agar plates (see Reagents below).

During its life cycle, the M13 does not lyse the JM107, but slows its replication. Bacteria that have lost their F factor cannot be infected by the M13, and have a selective growth advantage over infected cells. In addition, M13 phage containing inserts replicate more slowly than parental phage. These facts have several consequences:

1. When plated, JM107 infected with M13 does not produce a clear plaque, but rather a more translucent depression representing slower bacterial growth.

2. M13 phage with larger inserts tend to form smaller plaques on the plate, resulting in an inverse relationship between insert and plaque size.

3. Following an overnight minigrowth of M13 in JM107, a significant portion of the bacteria may have lost its F factor. If this growth is used to inoculate a large scale M13 growth, the phage yield may be low. Therefore, when doing a large scale M13 growth, a fresh (uninfected) overnight growth of JM107 should be used, together with the supernatant (containing filamentous phage but no bacteria) of the M13. This approach maximizes M13 yield by avoiding significant contamination with F- bacteria, which have a selective growth advantage over the F+ M13-infected bacteria.

4. Phage containing inserts may be contaminated with a small amount of parental M13 which is undetectable in overnight minigrowths, but which form a significant portion of the total phage following a large scale growth. For this reason it is important to purify recombinant phages by plating at low titer and picking a single plaque before attempting large scale growths.

Below is a description of several strategies for subcloning DNA into M13 and for sequencing the insert using the dideoxy chain-termination method.

SUBCLONING INTO M13

Preparing RF M13
The filamentous phage M13mp18 and M13mp19 and the uninfected JM107 can be readily obtained from commercial sources. Uninfected JM107 and M13-

containing bacteria are grown overnight at 37°C in 10 ml of 2 × YT (see Reagents below). The following morning, 9 ml of the uninfected JM107 and 4.5 ml of the supernatant from the M13 growth are added to 1 liter of 2 × YT and grown for 7 hours at 37°C in a shaking incubator. Double stranded phage DNA is prepared from the bacterial pellet by the alkali lysis procedure (see reference[2] and protocol at end of chapter) until after the isopropanol step. The isopropanol DNA pellet is resuspended in 8 ml of TE, 4 ml of 3 × high salt buffer (see Reagents below) are then added, and the solution is chilled at 4°C for 1 hour to precipitate high molecular weight RNA. The RNA is removed by centrifugation at 10 000 rpm for 10 minutes at 4°C and phage DNA is precipitated from the supernatant with 2 volumes of 95% ethanol at −20°C overnight.

The following day, the DNA is pelleted, washed with 70% ethanol and lyophilized dry. Double stranded phage DNA is purified by banding following centrifugation on a CsCl gradient. This is done by resuspending the DNA pellet in 4.8 ml of TE, followed by the addition of 8.4 g of CsCl and 0.8 ml of ethidium bromide (10 mg/ml). 8 ml of a CsCl solution (10.2 g/16 ml of TE) are placed into each of two 17 × 76 mm Beckman Quickseal polyallomer tubes, and half of the DNA-CsCl solution is layered underneath using a Pasteur pipet. The tubes are sealed and centrifuged at 65 000 rpm for 5 hours at 20°C in a Type 75 TI rotor with the brake off. After centrifugation, a single band should be visible at approximately the middle of the tube. The band is removed by piercing the side of the tube with a needle.[2] Ethidium bromide is extracted with either iso- or n-butanol. The aqueous phase is increased to 3 times the original volume, and the DNA is precipitated with 6 volumes of 95% ethanol overnight at −20°C. Aliquots of the DNA can now be pelleted, washed with 70% ethanol to remove any remaining CsCl, and digested with appropriate restriction endonuclease enzymes.

Inserting DNA into M13

The appropriately linearized RF M13 is ligated with the DNA insert at a 1 : 1 molar ratio.[2] Ligation reactions include 150–200 ng of M13 in a 15 μl reaction using T_4 DNA ligase. A control ligation with linearized M13 alone is done in parallel. The ligation is done at 12°C overnight. Concurrently, 10 ml of JM107 in 2 × YT are grown overnight at 37°C. The following morning, 0.5ml of the overnight stationary growth of JM107 is added to 50 ml of 2 × YT in a 250 ml flask and incubated for approximately 2 hours until the bacteria is in log growth (OD_{600} = 0.3–0.6). The bacteria are pelleted, resuspended in 20 ml of cold sterile 50 mM $CaCl_2$ and incubated at 4°C for 20 minutes. They are then repelleted and resuspended in 1 ml of 50 mM $CaCl_2$. 200 μl of the $CaCl_2$-treated bacterial suspension are added to each 15 μl ligation reaction and incubated at 4°C for 30 minutes. The cells are then heat shocked at 42°C for 2 minutes and added to a mixture of 200 μl of the remaining overnight JM107, 10 μl of 100 mM IPTG, and 50 μl of a 0.2% XGAL solution in N,N-dimethylformamide. 3 ml of top agar (see Reagents) are added and the mix is plated on a small YT agar (see Reagents) plate. With incubation at 37°C, plaques are visible after approximately 3 hours and blue coloration by 6 hours. On the control plate (ligated linearized M13 only), background clear plaques should comprise approximately 0.5% of the total plaques, and a successful ligation should yield 1–10% clear plaques.

Detecting and characterizing positive clones
Clear plaques can be grown by inoculating the plaque into 10 ml of $2 \times$ YT medium and incubating with aeration overnight at 37°C. Alkali lysis minipreparations are performed on 1–1.5 ml of the overnight growths.[2] Clones containing the insert are identified after digestion with the appropriate restriction enzyme(s) and fractionation on agarose gel.

Single stranded DNA (ssDNA) is prepared from 8 ml of the overnight growth of the positive plaques to determine the orientation of the inserts relative to the M13 DNA. Bacteria are pelleted and the supernatant containing the filamentous phage is removed. $\frac{1}{20}$ volume of both 40% PEG (MW 6000) and 5 M sodium acetate, pH 7.0, are added to the supernatant, and the filamentous phage is precipitated at 4°C for at least one hour and then pelleted at 10 000 g for 15 minutes. The tubes are inverted to completely drain the PEG solution, and excess liquid is wiped away with either a Kimwipe or a cotton swab, carefully avoiding the pellet. The drained pellet is resuspended in 500 μl of TE and extracted successively with an equal volume of phenol, 1 M Tris HCl, pH 8; phenol : chloroform (1 : 1); and then chloroform. The ssDNA is then precipitated at -70°C for 15–30 minutes following the addition of NaCl to 0.25 M and two volumes of 95% ethanol. The DNA is pelleted, washed with 70% ethanol, dried and resuspended in 50 μl of TE.

To determine relative insert orientation in the positive clones, heteroduplexes between pairs of clones are formed followed by S_1-nuclease digestion and electrophoresis on an agarose gel (Fig. 9.3). Clones with inserts in opposite orientations will anneal to each other, and thus contain an S_1-nuclease resistant double stranded DNA fragment equal in size to the full-length insert. Procedures for heteroduplex formation and analysis are as follows: 4.5 μl of each of two of ssDNA preparations from two independent clones are mixed with 1 μl of 10 $\times$ NaTMS (see Reagents), and then placed in a heating block at 100°C. The heat supply to the block is turned off, and when the block temperature falls below 45°C, 12.5 μl of $4 \times$ S_1-nuclease buffer (see Reagents) and 3 units of S_1-nuclease are added to a final reaction volume of 50 μl. The reactions are incubated for 15 minutes at 37°C and then electrophoresed on an agarose gel.

After two clones are found with the insert in opposite orientations, the remaining 0.5 ml of their minigrowths are used to obtain purified clones by dilutional plating and selection of isolated plaques. Purified plaques are saved as toothpick streaks on YT agar plates, and supernatants of minigrowths from these plaques can be used for long-term storage at -20°C. These purified clones can be used for the preparation ssDNA for sequencing as described below.

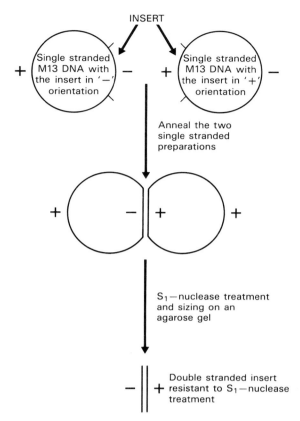

Fig. 9.3 Heteroduplex formation between two single stranded M13 phage with inserts followed by digestion with S_1 nuclease. The complementary inserts form a region of double stranded DNA resistant to S_1 nuclease digestion.

M13-DIDEOXY SEQUENCING

Preparing the single stranded template for sequencing

5 ml of 2 × YT in a 50 ml sterile plastic centrifuge tube are inoculated with the purified clone, while a second tube is inoculated with uninfected JM107. Both tubes are grown overnight at 37°C, and in the morning; 50 µl of the JM107 culture are inoculated into 12 ml of 2 × YT in a 125 ml Bellco flask and incubated with shaking (200 rpm) for an hour at 37°C. 50 µl of the supernatant from the overnight M13 growth are then added to the flask and incubated for 8–12 hours. The ssDNA is then prepared as described above, yielding approximately 25–50 µg for sequencing.

Sequencing reactions
In a 400 μl siliconized microfuge tube add:
  2 μl (approximately 1 μg or ⅓ pmol) M13 template
  2 μl (2 ng or ⅓ pmol) 15 nucleotide primer
  2μl 10 × NaTMS
  8 μl H₂O
  totalling 14 μl in a 100°C heating block which is turned off and allowed to cool to less than 45°C.
While the DNA is annealing prepare 4 400 μl siliconized tubes:
  Tube A: 1 μl A° mix and 1 μl ddA (0.25 mm)
  Tube C: 1 μl C° mix and 1 μl ddC (0.5 mm)
  Tube G: 1 μl G° mix and 1 μl ddG (0.25 mm)
  Tube T: 1 μl T° mix and 1 μl ddT (1.0 mm)
(Recipes for mixes in Reagents. The ddNTP's concentrations may have to be varied if the reactions terminate too early or late).

The tube containing the annealed DNA is centrifuged, placed on ice, and 1 μl (10 μCi) of α[^{32}P] or α[^{35}S] dATP) and 1 μl (1 unit) of Klenow added. Mix the contents by pipeting in and out of a micropipet tip. 3 μl of the annealed DNA mix are added to each of the four tubes containing dNTPs and ddNTPs, and incubated at 37°C for 15 minutes. Then 1 μl of the chase mix (see Reagents) is added, and the tubes are incubated at 37°C for another 15 minutes. Reactions are terminated by the addition of 10 μl of the formamide dye mix (see Reagents) and then heated to 100°C for 3 minutes before applying 2–3 μl of each reaction to a denaturing polyacrylamide gel. Details on how to make and run the sequencing gel are given in Chapter 11, Part One.

Very recently, we have begun to use the modified T₇ polymerase (Sequenase®) sequencing kit (United States Biochemical Co., Cleveland, Ohio). We follow manufacturer's instructions except that after adding the Sequenase to the labeling reaction 4 μl are removed at 0.5, 1.0, 2.0, 5.0 and 10.0 minutes and collected into a single eppendorf tube on ice. The termination reaction is then carried out as described by the manufacturer. This new sequencing technique yields cleaner sequences that can be read further than reactions using Klenow. The modification described above yields more uniform labeling over the readable sequence, and we now can routinely read 400–500 bp for each sequence reaction.

Sequencing strategies
As already noted, the major limitation of M13-dideoxy sequencing is the inability to read a DNA sequence more than approximately 300 bp downstream from the universal primer site. Unless the insert is small, determination of the full-length sequence therefore requires the development of a sequencing strategy. I will now briefly describe several approaches we have used alone or in combination for sequence determination of longer inserts. Major advantages and disadvantages of each approach are discussed, and references are included that describe each approach in more detail.

One strategy uses Bal 31 exonuclease to obtain an ordered set of overlapping deletion mutants for sequence analysis (see Fig. 9.4). A double stranded insert within a circular phage or plasmid is linearized at a unique restriction site near one end of the insert. A timed Bal 31 exonuclease digest is done as follows:

2.5 µl linearized DNA (1 µg/µl)
7.5 µl 4 × Bal 31 buffer
1 µl Bal 31 (1 U/µl)
19 µl H$_2$O

A pilot digest is done to determine the rate of Bal 31 digestion. Aliquots are collected at 5–10 minute intervals, and these aliquots are run on an agarose gel. The Bal 31 reaction is repeated, and appropriate aliquots are pooled, phenol extracted and ethanol precipitated. An endonuclease reaction that uniquely cuts near the opposite end of the insert is then done, resulting in a family of fragments with random Bal 31 (presumably blunt) ends and a fixed unique restriction point at the other end. The family of insert fragments can be separated and purified away from the phage fragments following electrophoresis and elution from an agarose gel. With many vectors the purification of insert fragments from the vector is unnecessary and the electrophoresis and elution steps can be omitted. These Bal 31-generated deletions are then ligated into an appropriately restricted M13 phage containing a blunt end near the universal primer and the unique restriction site at the other end. The ligated M13 is then transfected into JM107. ssDNA is isolated from individual clones and annealed to single stranded M13 containing the complementary strand of the full-length insert, and the duplexes are then treated with S$_1$-nuclease and resistant heteroduplexes sized as described above.

One major advantage of this strategy is that digestion with Bal 31 can be carefully timed. The degree of digestion can thus be regulated to target sequencing of a specific subregion of the insert. Major disadvantages of this strategy are that restriction map information about the insert and phage or plasmid is required and that the inserts have to be religated into M13.

Another rapid DNA subcloning strategy uses T$_4$ DNA polymerase to digest linearized single stranded M13 (see Fig. 9.5). ssDNA is annealed to an appropriate oligonucleotide primer and the duplex portion is then linearized by digestion with the appropriate restriction enzyme. The 3' to 5' exonuclease activity of T$_4$ DNA polymerase is used to progressively shorten the insert, leaving the universal primer site intact. The ssDNAs are tailed by the addition of dATP (for M13mp18) or dGTP (for M13mp19). The tailed ssDNA is then annealed to an oligonucleotide primer containing a complementary tail. After ligation, circularized ssDNA is transfected into JM107, and the insert size of phage from clear plaques can be determined as described above. Alternatively, a more rapid but less accurate technique can be used for sizing inserts. This procedure is as follows: 1 µl of 2% SDS and 3 µl of loading buffer are added to 20 µl of each supernatant from overnight growths, and the samples are then sized on a 1.0% agarose gel following electrophoresis overnight at 100 V.

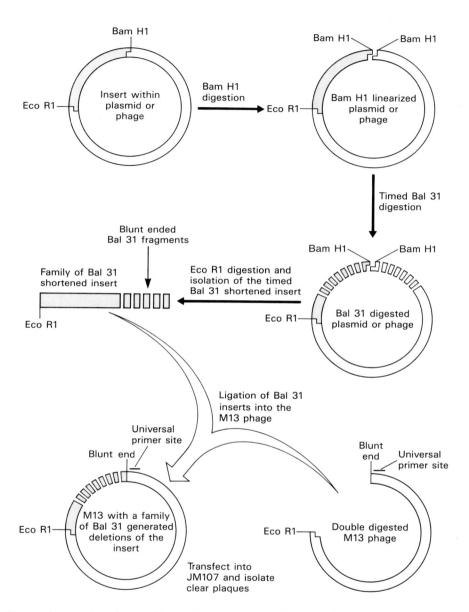

Fig. 9.4 Construction of a set of insert deletions using Bal 31 digest of the linearized double stranded phage or plasmid with insert. The dashed region refers to the variably digested DNA by the Bal 31 which are blunt ended and inserted near the universal primer site of the M13 vector.

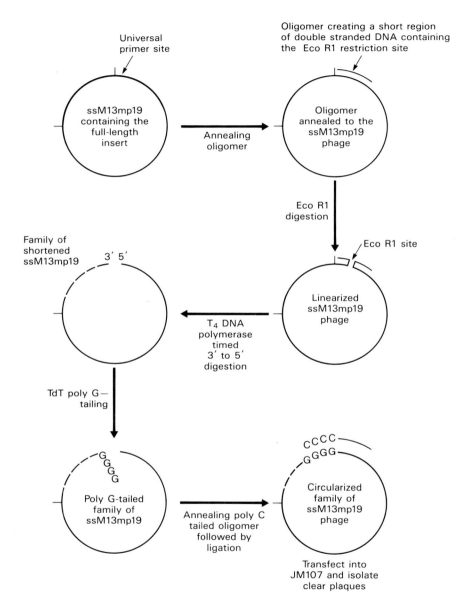

Fig. 9.5 Construction of a set of insert deletions using T_4 DNA polymerase digest of linearized single stranded M13mp19 containing the insert. The dashed region refers to the variably digested DNA by the T_4 DNA polymerase that is then re-annealed to the universal primer site of the M13.

The detailed description below is essentially as described by Dale et al.[3] The first step in this procedure is to anneal the appropriate commercial oligonucleotide to the M13 used as follows:

    1–4 μl ssM13 with insert (1 μg/μl)
    2 μl 10 × Oligo buffer
    1 μl commercial oligomer (0.5 A260/ml)
    14 μl H$_2$O

This is placed in a 100°C heating block which is then turned off. When the temperature is less than 40°C, 10 U of Eco RI (for M13mp19) or of Hind III (for M13mp18) are added and the digestion is allowed to proceed overnight at 42°C. The next morning the linearized DNA is treated with T$_4$ DNA polymerase to create a family of deletions:

    20 μl linearized ssM13
    2 μl 10 × T$_4$ DNA polymerase buffer
    2.5 μl 100 mM DTT
    1 μl BSA (10 mg/ml)
    1–2 U of T$_4$ DNA polymerase

Aliquots are collected every 5–15 minutes for 45–60 minutes, pooled and heat inactivated at 65°C for 10 minutes. The ssM13 DNA fragments are then dATP tailed (for M13mp18) or dGTP tailed (for M13mp19) using TdT (terminal transferase) as follows:

    3 μl appropriate dNTP
    4 U TdT

The reaction is incubated at 37°C for more than 10 minutes. 10 μl of the reaction are then removed and saved as a unligated control. 1 μl of the same oligomer added above is again added and re-annealed as described above. 3 μl of 10 mM ATP, 4 μl of H$_2$O, and 1–3 U of T$_4$ DNA ligase are added and ligation is allowed to proceed overnight followed by transfection into JM107.

Two major advantages of the T$_4$ DNA polymerase approach are that restriction map information is not necessary and that the insert is ligated in an intermolecular reaction to M13. In addition, the necessary oligomers for this approach are commercially available. T$_4$ DNA polymerase does, however, tend to generate a wide spectrum of differently-sized inserts at all time points, making this approach less useful for targeting sequence determination to a particular subregion of the insert. Many clones may therefore have to be sized to find those within the size range of interest.

Not all approaches for rapid dideoxy sequencing involve creation of deletion mutants to bring more distal sequences into juxtaposition with the M13 universal primer site. One other method relies on the synthesis of 12–18 nucleotide (n) primers which are complementary to the single stranded insert at 200–300 n intervals along the region of interest. This approach has several limitations in that it requires use of a synthetic oligonucleotide constructed either manually or by automated sequencer. In addition, we have had variable results from such primers, probably related to primer sequence and to the quality of the oligonucleotide synthesis and purification. Since synthesis of each oligomer yields sufficient material for thousands of reactions, this approach is probably best suited for situations in which sequence information is required from the same DNA from different independent sources (for example, sequencing many β-globin genes from many thalassemic patients). Synthetic oligonucleotide primers can also be useful for targeted sequencing of regions not amenable to analysis by construction or deletions. In this type of situation, the oligomer can be exactly tailored to determine the sequence of a region not attainable by the above two strategies.

We sometimes subclone restriction digest fragments of a full-length insert to obtain the sequence of a region of interest. M13mp18 and 19 with their rich MCS offer the versatility of subcloning most single or double restricted fragments into M13. While attempts to sequence an entire insert based on subcloning restriction enzyme subfragments has had only limited success in our hands, this approach should be kept in mind as an alternative solution for sequencing short stretches of DNA missed by either the Bal 31 or T_4 DNA polymerase strategies.

REAGENTS

2 × YT (per liter):
  16 g tryptone
  10 g yeast extract
  5 g NaCl

Minimal agar (per liter):
  15 g Bacto-agar
  10.5 g $K_2 HPO_4$
  4.5 g $KH_2 PO_4$
  1 g $(NH_4)_2SO_4$
  0.5 g sodium citrate
  0.2 g $MgSO_4 7H_2O$
  5 μg thiamine HCl
  2 g glucose

(Except for the bacto-agar, these items are sterilized separately as a concentrated solution.)

4 × S_1 Nuclease buffer:
  200 mM sodium acetate, pH 4.5
  600 mM NaCl
  2 mM $ZnSO_4$

  A° mix: 20 μl 0.5 mM dTTP
            20 μl 0.5 mM dCTP
            20 μl 0.5 mM dGTP
            20 μl 10 × NaTMS

$C°$ mix: 20 μl 0.5 mM dTTP
              1 μl 0.5 mM dCTP
              20 μl 0.5 m dGTP
              20 μl 10 × NaTMS
$G°$mix: 20 μl 0.5 mM dCTP
              20 μl 0.5 mM dCTP
              1 μl 0.5 mM dGTP
              20 μl 10 × NaTMS
$T°$ mix: 1 μl 0.5 mM dTTP
              20 μl 0.5 mM dCTP
              20 μl 0.5 mM dGTP
              20 μl 10 × NaTMS

(The exact ratio of dNTP/ddNTP may have to be varied to maximize sequencing and avoid excessive early or late terminations.)

Chase mix:
    0.4 mM of all 4 dNTPs

Formamide dye:
    0.1% (w/v) xylene cyanol
    0.1% (w/v) bromophenol blue
    10 mM Na_2 EDTA
    95% (v/v) deionized formamide
    (store at 4°C)

4 × Bal 31 buffer:
    48 mM $CaCl_2$
    48 mM $MgCl_2$
    2.4 M NaCl
    80 mM Tris HCl, pH 8.1
    4 mM EDTA

10 × Na TMS:
    70 mM Tris HCl, pH 7.5
    70 mM $MgCl_2$
    500 mM NaCl

10 × Oligo buffer:
    100 mM Tris HCl, pH 7.4
    100 mM $MgCl_2$

10 × T_4 DNA polymerase buffer:
    330 mM Tris acetate, pH 7.8
    660 mM K acetate
    100 mM Mg acetate

REFERENCES

1 Messing J. New M13 vectors for cloning. Methods Enzymol 1983; 101: 20–78.
2 Maniatis T, Fritsch E F, Sambrook J. Molecular cloning — a laboratory manual. New York: Cold Spring Harbor Laboratory, 1982.
3 Dale R M K, McClure B A, Houchins J P. A rapid single stranded cloning strategy for producing a sequential series of overlapping clones for use in DNA sequencing: Application to sequencing the corn mitochondrial 18s rDNA. Plasmid 1985; 13: 31–40.
4 Poncz M, Solowiejczyk D, Ballantine M, Schwartz E, Surrey S. 'Nonrandom' DNA sequence analysis in bacteriophage M13 by the dideoxy chain-termination method. Proc Natl Acad Sci USA 1982; 79: 4298–4302.

10
Detection of mRNA in frozen tissue sections by *in situ* hybridization with RNA probes

J. W. Schneider   M. F. Utset

INTRODUCTION

A powerful method for studying the expression of cloned genes is hybridization of nucleic acid probes to mRNA sequences in tissue sections. For example, *in situ* hybridization of tissue sections cut from whole fetal rats allows the anatomic distribution of specific mRNAs to be determined, without the formidable task of preparing RNA samples from each tissue. The sensitivity and selectivity of *in situ* hybridization matches that of filter blotting techniques. In addition, the regional or cellular site of mRNA expression can be localized within complex tissues comprised of heterogeneous cell types. *In situ* hybridization was originally developed as a method of mapping repetitive DNA sequences to the giant polytene chromosomes of Drosophila; indeed, this method is now routinely used for the localization of genes to mammalian metaphase chromosomes. Use of *in situ* methods for analysis of mRNA has been developed more recently. One of the earliest adaptations of the technique to RNA detection was the study of globin mRNA expression in erythroid cells of developing fetal liver.[1]

Histologic specimens contain high concentrations of RNA molecules. RNA is the major basophilic staining (blue) component of cytoplasm. *In situ* hybridization of tissue sections using radiolabeled nucleic acid probes permits low abundance messages to be specifically detected in this complex mixture of cellular RNAs. The sensitivity of this technique has been estimated to be high enough to detect an mRNA species which represents ~0.01% of the target cell's total message (i.e. an mRNA present at only ~100 copies per cell).[2] Gentle tissue fixation, hybridization, and washing conditions have been developed which permit these low abundance mRNAs to be detected while preserving the histological architecture of the section.

In this chapter, we describe a simple and reliable *in situ* hybridization protocol which can detect the expression of specific mRNAs in the various tissues of a fetal rat. This method is particularly useful for studying the differential expression of highly homologous protein isoforms, where primary structural similarities make identification at the protein or immunologic levels difficult. In such cases, the mRNAs encoding these protein isoforms may be readily distinguished by high stringency nucleic acid hybridization. In addition, the 5'- or 3'-untranslated sequences of each isoform mRNA may be totally unrelated and therefore provide a unique molecular probe for each RNA species. We have used the procedures presented here to characterize the distribution patterns of Na,K-ATPase mRNAs in fetal rats (Fig. 10.2a, b), and of homoeo box mRNAs in the central nervous system of newborn mice.[3] However, this technique should be readily applicable to the study of virtually any mRNA expressed at levels of > 100 copies/cell in most tissues or cell blocks.

The procedure requires that several independent components be prepared separately prior to the hybridization reactions: glass slides, tissue sections, and radiolabeled probe. The preparation of these components is described first.

PRE-TREATMENT OF GLASS SLIDES

We have found that frozen tissue sections adhere well to acid-washed and gelatin-coated glass microscope slides. These slides also have low background due to non-specific binding of the probe to the glass. Wash the slides first in warm laboratory detergent, then rinse thoroughly with distilled water. Acid treat slides for ~30 minutes in 1% (v/v) hydrochloric acid, then rinse again with distilled water. Dip slides several times in 0.3% gelatin subbing solution, made by dissolving 1.5 g gelatin (Macalaster Bicknell Co.) in 500 ml water with heat. This solution can be stored for 3–4 weeks at 4°C. Air dry the slides, which can be stored indefinitely in a dust-free container at room temperature. Several references have also described the use of transparent tape rather than glass slides for collecting and processing the tissue sections.[4,5]

PREPARATION OF FROZEN TISSUE SECTIONS

A pregnant rat at ~14–18 days gestation is anesthetized by ether and killed by decapitation. The uterus is removed and cooled on ice in phosphate buffered saline (PBS, see formula below). The fetuses are dissected free of fetal membranes and maternal tissues, and immediately frozen in OCT (a water soluble embedding medium, Tissue Tek II, Miles Laboratories) on a specimen holder by immersing the holder in liquid nitrogen or dry ice. Once the fetus is completely embedded and frozen, the block is placed in the −20°C cryostat chamber for at least 20 minutes to equilibrate to blade temperature. Frozen sections of the fetus, 4–20 μm thick, are then cut with the −20°C microtome and collected onto pre-treated glass slides. Generally, thinner sections are retained better on the slide and also yield better tissue morphology. However, it does require practice to cut a very thin section of a large fetus and transfer it intact to a microscope slide. Carefully flatten the sections onto the slide with a

Fig. 10.1 (opposite) *In situ* hybridization to mRNA in frozen tissue sections. A frozen rat fetus is sectioned and the slices mounted onto pre-treated microscope slides. The slides are fixed with paraformaldehyde and then hybridized with ^{35}S-radiolabeled RNA probe (riboprobe) to determine presence of homologous mRNA. The hybridized sections are washed, treated with RNase A, and then directly autoradiographed on film. The slides can be stained and compared with the corresponding autoradiograph for identification of positive hybridization signals.

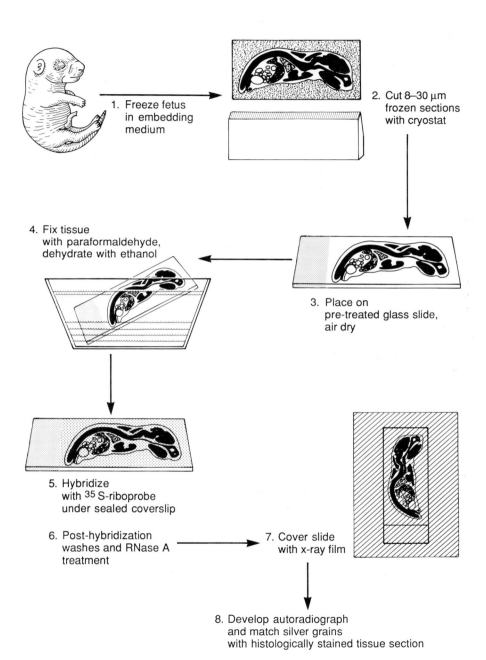

soft bristle paintbrush while warming the underside with your finger. Several sections may be placed on each slide and the remaining fetal tissue can be recoated with OCT and saved at −70°C for later use. The slides are placed on a 50°C hot plate for 2 minutes, then air dried for several hours.

Several published protocols include a pre-fixation step during which the tissue section is permeabilized by brief digestion with proteinase K[2,6]. Others, including us (see below), have found no advantage to this proteolytic treatment for obtaining strong hybridization signals.[7] When the tissue section is a whole fetus, the appropriate conditions for proteinase K treatment must be determined empirically, since different tissues within the section tolerate this enzyme to varying extents. If inadequate penetration of the probe to mRNA within the tissue is a concern, it is advisable to try cutting thinner sections or using a short length probe before including proteinase K digestion.

In order for cellular mRNAs to be available for hybridization *in situ*, they must be fixed to the cytoplasmic matrix with their nucleotide sequence exposed and available for base pairing with the complementary probe. Several careful studies have established that 4% paraformaldehyde is superior to all other fixatives for cross-linking RNA within the tissue, while preserving cellular detail.[2,7] We place the dry tissue section slides in freshly prepared fixative, made by dissolving 20 g paraformaldehyde (Aldrich Chemical Co.) in 500 ml of phosphate buffered saline (PBS, $NaH_2PO_4H_2O$, 1.38 g/l, and NaCl, 9.0 g/l, adjust to pH 7.5 with 1 N NaOH), for 20 minutes at room temperature. The slides are rinsed first for 5 minutes in 3 × PBS, and then twice for 5 minutes each in 1 × PBS. The tissue sections are then dehydrated by a series of ethanol washes for 2 minutes each: 30% EtOH, 60% EtOH, 80% EtOH, 95% EtOH, and finally 100% EtOH. The slides are allowed to air dry and stored in a desiccated plastic box at −20°C. (Editor's note: blood cell or tissue culture cell suspensions can be pipeted and then mixed directly as a dispersed suspension in paraformaldehyde.)

An alternative method, if an individual rat organ such as the brain is to be analyzed, is the perfusion of the animal with fixative before dissection. Perfusion of the rat will result in the best preservation of both RNA and tissue. The rat should be perfused transcardially, first with cold PBS to wash out the blood and then with 4% paraformaldehyde in PBS for ~20 minutes, essentially as described in reference.[10] The brain (or other organ) is removed, sliced into appropriate fragments and post-fixed for 1 hour to overnight in 4% paraformaldehyde in PBS at 4°C. The tissue may also be soaked overnight in 15% sucrose in PBS at 4°C. This treatment prevents the formation of ice crystals during freezing. The tissue can then be frozen in OCT, sectioned, and processed as described above.

PREPARATION OF ^{35}S-LABELED RNA PROBES

There are several advantages to using [^{35}S]thio-substituted RNA probes (riboprobes) for *in situ* hybridization experiments. First, [^{35}S]ribonucleotides are more convenient to work with at the laboratory bench than [^{32}P]-labeled nucleotides.

The energy of β-emission from ^{35}S generates signals which can be detected by either direct autoradiography on film (for regional localization) or by dipping slides in photographic emulsion (for cellular localization). Second, high specific activity ($\sim 10^8$ cpm/μg) riboprobes can be reliably produced by *in vitro* RNA transcription of a cDNA sequence after subcloning into a Gemini expression vector (Promega Biotec). Futhermore, with the Gemini vector both strands of the cDNA can be transcribed by either Sp6 or T_7 RNA polymerase; the labeled 'antisense' transcript hybridizes to mRNA while the 'sense' transcript provides an excellent hybridization control for non-specific background. A further advantage to using riboprobe is that a post-hybridization RNase A wash step will digest most non-hybridized probe molecules and thereby greatly reduce background.

The RNA transcriptions are set up with reagents provided in the Riboprobe Gemini II System kit (Promega Biotec) in the following reaction mix (complete instructions are provided with the kit; see also Ch. 4):

12 μl 5 × transcription buffer (200 mM Tris HCl, pH 7.5, 30 mM $MgCl_2$, 10 mM spermidine, 50 mM NaCl)
6 μl 100 mM DTT
1.5 μl RNasin (25 U/μl stock; final concentration ~ 1 U/μl)
12 μl cold ribonucleotide mix (mix together equal amounts of 10 mM each rATP, rCTP, rGTP and water; mix contains 2.5 mM each rATP, rCTP and rGTP)
4 μl cold 250 μM rUTP
2 μl linearized template DNA (~ 2 μg, prepared by restriction digestion, followed by phenol/chloroform extraction and ethanol precipitation)
18 μl [^{35}S]UTPαS (> 1000 Ci/mM, Amersham, SJ.1303)
2.5 μl H_2O
2 μl appropriate RNA polymerase, Sp6 or T_7, 3–30 U/μl (Promega Biotec or Boehringer Mannheim)
60 μl total volume

The RNA synthesis reaction is incubated for 1 hour at 37°C, after which 1 U RQl DNAase (Promega Biotec)/μg template DNA is added and the reaction mix incubated an additional 15 minutes at 37°C to destroy the template DNA. The probe is then separated from unincorporated ribonucleotides by passage through a Sephadex-G50 spin column. The procedure for spin column chromatography is described in detail in reference[8].

A debated parameter of *in situ* hybridization methodology is whether probe fragment length is important. Several authors suggest that small probe fragments (or synthetic oligonucleotides) 50–200 bases in length can better penetrate the tissue to locate their target mRNAs.[9] However, others[7] have demonstrated that large probe fragments > 1.5 kb are optimal for obtaining high signals with paraformaldehyde-fixed tissue sections. We have obtained strong hybridization signals with intact riboprobes 2–3 kb in length. Reference[9] describes a protocol for adjusting probe length by partial alkaline hydrolysis of the RNA, but we have not yet found this additional manipulation to be necessary.

HYBRIDIZATION OF TISSUE SECTION SLIDES

We have not found it necessary to include a pre-hybridization washing step for our *in situ* hybridization protocol. The hybridization buffer used with riboprobes has the following final composition:

1 × Denhardt's [0.02% each bovine serum albumin (Boehringer Mannheim), Ficoll (Sigma), and polyvinylpyrrolidone (Sigma)]
500 µg yeast transfer RNA (Sigma)/ml
0.75 M NaCl
0.01 M Tris HCl, pH 7.5
0.002 M EDTA
0.01 M dithiothreitol (Calbiochem)
50% deionized formamide (Boehringer Mannheim)

Approximately 5×10^7 cpms of the probe that has been recovered from the spin column are added to each ml of hybridization cocktail. Riboprobe is single strand RNA and does not have to be denatured before addition to the hybridization mix. 50 µl of hybridization solution is applied to each slide and gently covered with a 40 × 24 mm coverslip. It is important that air bubbles trapped under the coverslip be removed. Carefully wedge a razor blade under the coverslip and lift it to allow the bubbles to escape. The coverslips are then sealed with fresh rubber cement, and the slides incubated in a 50°C oven for 4 hours to overnight.

POST-HYBRIDIZATION WASHING SCHEDULE

The sections are washed by first removing the rubber cement and allowing the coverslips to float off in 50°C wash solution [2 × SSC (20 × SSC = 3.0 M NaCl, 0.3 M Na citrate, made by dissolving 175.2 g NaCl and 85.7 g Na citrate in 1 liter H$_2$O), 50% formamide, and 0.1% β-mercaptoethanol]. If the coverslips do not float off, gently wedge a razor blade under a corner while the slide is submerged in wash buffer.

The slides are washed at 50°C in fresh 2 × SSC, 50% formamide, and 0.1% β-mercaptoethanol for > 30 minutes. The sections are then incubated for at least 30 minutes at 37°C in RNase A buffer (0.5 M NaCl, 0.01 M Tris, pH 8.0, and 20 µg/ml RNase A, Sigma). RNase treatment is followed by a second 2 × SSC, 50% formamide, 0.1% β-mercaptoethanol wash for > 30 minutes at 50–60°C.

Fig. 10.2 (opposite) Detection of Na,K-ATPase α subunit mRNA in embryonic rat tissues using *in situ* hybridization. Panel a is a histologically stained para-sagittal section of an 18-day-old embryonic rat. Panel b is a direct autoradiograph of this same section after hybridization with ^{35}S-labeled rat Na,K-ATPase α subunit riboprobe. Positive hybridization signals (dark areas) were obtained in several fetal organ systems, including the kidney, gastrointestinal tract, nasal respiratory epithelium, and dorsal root ganglia. Panel c is a high magnification light field photomicrograph of a similar section after hybridization with the Na,K-ATPase α subunit probe followed by liquid emulsion autoradiography. Silver grains can be seen in the fetal kidney (Ki), fetal adrenal cortex (Ad) and dorsal root ganglia (Dg). Panel d is a dark field photomicrograph of this same region. In d, the bright areas represent positive hybridization signal.

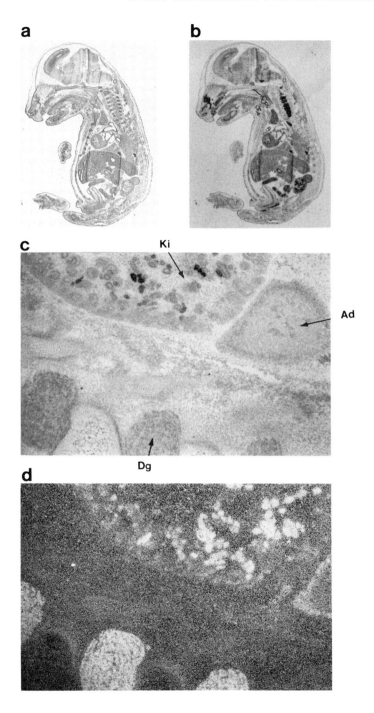

The temperature may be adjusted depending on the desired washing stringency. The final wash is 0.1 × SSC, and 0.1% β-mercaptoethanol for > 30 minutes at 50–60°C. The slides are dehydrated by two washes in 75% ethanol for 2 minutes each, followed by two washes in 95% ethanol, and then air dried.

AUTORADIOGRAPHY AND HISTOLOGIC STAINING

[^{35}S]Riboprobe hybridization signals can be detected by either direct film or liquid emulsion autoradiography. After exposure of the slide to a high resolution film such as Hyperfilm-βmax (Amersham), the autoradiograph can be examined side-by-side with the histologically stained section. This approach permits regional localization of positive signals. Alternatively, for resolution at the cellular level, the slide can be dipped in a liquid photographic emulsion (NTB-2, Kodak). This technique is described in detail in reference[11]. The slides can be stained by Giemsa or hematoxylin and eosin, as described in standard histochemistry texts. Reference[12] is a useful atlas for identifying anatomical structures in whole body sagittal sections.

This chapter describes a method of *in situ* hybridization that can be used to localize specific mRNAs in frozen sections of whole fetal rats. This technique can be used to detect most mRNAs for which cloned probes are available. The only specialized equipment required for this procedure that is not generally found in a molecular biology laboratory, is a cryostat. If you are not familiar with the operation of a cryostat, we suggest that you befriend a helpful colleague from pathology or cell biology, for whom frozen sectioning may be a daily routine.

REFERENCES

1 Harrison P R, Conkie D, Paul J, Jones K. Localisation of cellular globin messenger RNA by *in situ* hybridization to complementary DNA. FEBS Lett 1973; 32: 109–112.
2 Hafen E, Levine M, Garber R L, Gehring W L. An improved *in situ* hybridization method for the detection of cellular RNAs in Drosophila tissue sections and its application for localizing transcripts of the homoeotic Antennapedia gene complex. EMBO J 1983; 2: 617–623.
3 Awgulewitsch A, Utset M F, Hart C P, McGinnis W, Ruddle F H. Spatial restriction in expression of a mouse homoeo box locus within the central nervous system. Nature 1986; 320: 328–335.
4 Rall L B, Scott J, Bell G I, Crawford R J, Penschow J D, Niall H D, Coghlan J P. Mouse prepro-epidermal growth factor synthesis by the kidney and other tissues. Nature 1985; 313: 228–231.
5 Kodelja V, Heisig M, Northemann W, Heinrich P C, Zimmermann W. α$_2$-Macroglobulin gene expression during rat development studied by *in situ* hybridization. EMBO J 1986; 5: 3151–3156.
6 Lewis S A, Cowan N J. Temporal expression of mouse glial fibrillary acidic protein mRNA studied by a rapid *in situ* hybridization procedure. J Neurochem 1985; 45: 913–919.
7 Lawrence J B, Singer R H. Quantitative analysis of *in situ* hybridization methods for the detection of actin gene expression. Nucleic Acids Res 1985; 13: 1777–1799.
8 Maniatis T, Fritsch E F, Sambrook J. Molecular cloning: A laboratory manual. New York: Cold Spring Harbor, 1982.

9 Cox K H, DeLeon D V, Angerer L M, Angerer R C. Detection of mRNAs in sea urchin embryos by *in situ* hybridization using asymmetric RNA probes. Dev Biol 1984; 101: 485–502.
10 Swanson L W, Sawchenko P E, Rivier J, Vale W W. Organization of ovine corticotropin-releasing factor immunoreactive cells and fibers in the rat brain: an immunohistochemical study. Endocrinology 1983; 36: 165–186.
11 Stein G H, Yanishevsky R. Autoradiography. In: Jakoby W B, Pastan I H, eds. Methods in Enzymology. Vol LVIII. Cell culture 1979: p 279–292.
12 Theiler K. The house mouse. New York: Springer-Verlag, 1972.

11
Oligonucleotide design and use; *in vitro* mutagenesis

PART ONE
Oligonucleotide-directed site-specific mutagenesis
S. J. Baserga

BACKGROUND

Over the past ten years, oligonucleotide-directed site-specific mutagenesis of cloned DNA molecules has developed into a technique which is both feasible and rapid. The uses of site-specific mutagenesis are many; in general, this approach allows the investigator to mutate a cloned DNA molecule *in vitro* for further studies at the DNA, RNA or protein level. There are three major improvements or innovations that have made this technology accessible to most research laboratories. First, the increasingly widespread availability of oligonucleotide synthesizing machines has made access to the necessary oligonucleotides much easier and less expensive. Second, the introduction of the filamentous phage, M13, as a template for the mutagenesis[1] has made the procedure more convenient. Third, the use of repair-deficient *E. coli* as the host strain for template production[2] has made the procedure more efficient. In this review I will discuss the development of our current techniques, focusing on practical knowledge, and then briefly discuss some of the alternative methods.

The first successful attempts at oligonucleotide-directed site-specific mutagenesis were reported in 1978, in two different laboratories.[3,4] In both cases they used the technique to mutate an endogenous gene in the filamentous phage, Phi X174. A significant advance was made when Zoller & Smith[5,6] described the use of M13-derived vectors for the mutagenesis. In 1985 Thomas Kunkel reported the use of an *E. coli* deficient in the DNA repair enzymes *dut* and *ung* as a host cell for the growth of the DNA template. The designations *dut* and *ung* stand for mutations in the dUTPase and uracil-N-glycosylase genes, respectively; they cause uracil to be incorporated and to accumulate instead of thymine. This creates a bias against the survival of this strand, thereby increasing the mutation frequency. In our laboratory we have used the *E. coli* strain developed by Thomas Kunkel and the procedure developed by Zoller & Smith, with some minor modifications; a combination of these two has been found to be the most efficient.

In general, the technique of site-specific mutagenesis involves the following (see Fig. 11.1). Single stranded DNA bearing the clone of interest is prepared and annealed to a short oligonucleotide primer. The primer is an exact complement of the DNA except where the mutation is desired. We have used DNA

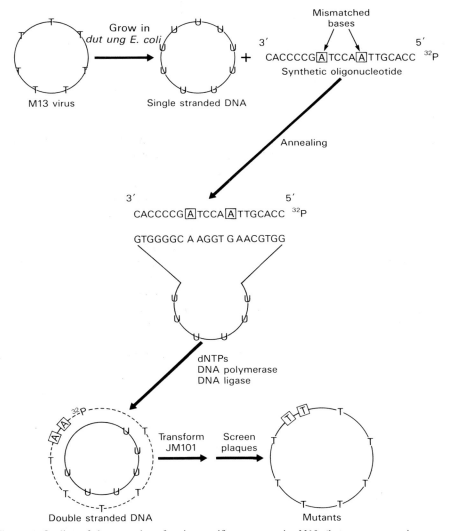

Fig. 11.1 Outline of the procedure for site-specific mutagenesis. M13 phage was grown in a repair-deficient *E. Coli* strain which allows uracil to be incorporated into the template DNA. The synthetic oligonucleotide is annealed to the template DNA and then extended in the presence of deoxynucleotides, T_4 DNA ligase, and DNA polymerase (Klenow). The reaction mixture is then transformed into JM101, a non-repair deficient host, and the mutants are identified.

cloned into M13 and grown in the *E. coli dut ung* strain. In the presence of deoxyribonucleotides, Klenow enzyme and T_4 DNA ligase, the complementary strand is synthesized. Upon transformation of *E. coli*, plaques are generated, some of which bear the mutation. At this step it is necessary to identify the mutant-bearing phage. A number of methods are available including biological assays, hybridization, restriction endonuclease digestion and direct DNA sequencing. With the current available methods, the mutation frequency is 40–80% and the clones can be screened by M13-dideoxy sequencing. Following identification of the correct plaque, replicative form (RF) of M13 is prepared, and the mutant DNA of interest can then be manipulated in its double stranded form. Please refer to Chapter 9 for information on the growth and maintenance of both the single stranded and double stranded forms of the M13 phage.

SELECTION OF THE OLIGONUCLEOTIDE

The selection of the oligonucleotide for use in site-specific mutagenesis is determined by the region of the DNA molecule of greatest interest to the scientist. However, there are other considerations to be taken into account when designing an oligonucleotide. For example, more efficient mutagenesis is achieved when the oligonucleotide anneals only to the region to be mutated and not to the flanking DNA. The probability of mispairing elsewhere can be minimized through the use of oligonucleotides at least 18–20 bases in length. Using one of the many DNA sequence comparison computer programs available, determination of the extent of mispairing can be accomplished and the oligonucleotide can be modified from this information. In addition, the placement of the mutation in the oligonucleotide affects the mutation frequency.[7,8] It is suggested that at least 6–7 bases flank the mispaired base on either side to ensure that the mutated base is not clipped out by the 3′ exonuclease activity still present in the Klenow fragment.

Once the oligonucleotide has been synthesized, specific priming on the M13 DNA template can be confirmed in two ways. The synthetic oligonucleotide can be used as the primer for dideoxy sequencing; if priming is specific an unambiguous DNA sequence can be read from the gel. The method for this can be found in Chapter 9. The second method involves the generation of a restriction fragment of predictable size after primer extension with Klenow and deoxynucleotides on template grown in a non-repair deficient *E. coli* strain. The oligonucleotide is kinased (5′ end labeled with polynucleotide kinase) with a small amount of $[^{32}P]\gamma$-ATP. If a radioactive restriction fragment of the correct predicted length is generated, the priming is specific.

SELECTION OF BACTERIAL STRAINS

The choice of which *dut ung E. coli* to use is influenced by the M13 vector which has been chosen. There are three *dut ung* strains available; their genotypes are listed in Table 11.1. BW313, described for use in the mutagenesis procedure by

Table 11.1 Genotypes of available *dut ung* strains.

BW313	HfrKl16 PO/45[*lysA* (61–61)], *dut1, ung1, thi1, relA1*[2]
CJ236	*dut1, ung1, thi1, relA1*/pCJ105 (Cm^r)[9]
RZ1032	as BW313, but Zbd-279::*Tn10, supE44*[9]

Thomas Kunkel[2], supports growth of M13 phages which do not require an amber suppressor (Mp1-2, MP18-19). However, some investigators have found that this strain is more difficult to work with than a more recently developed strain, CJ236. This strain, described by Catherine Joyce at Yale[9] supports growth of M13 phages Mp1-2 and MP18-19. CJ236 has the advantage that chloramphenicol resistance is a selectable marker for its F factor. The F factor is necessary for infectivity of the M13 phage. A third strain, RZ1032, supports the growth of M13 phages Mp7-11 which require an amber suppressor to grow. There is no selectable marker for the presence of the F factor in this strain, and the strain loses its F factor at a low but detectable rate. It would probably benefit the investigator to subclone the DNA to be mutagenized into M13 Mp1-2 or Mp18-19 and to use the CJ236 as the host strain for template production. A more detailed account of the growth and maintenance of *dut ung* strains can be found in reference 9.

MATERIALS

Bacterial strains:
appropriate *dut ung E. coli* strain (see Table 11.1)
any *dut$^+$ ung$^+$ E. coli* strain (e.g. JM101)

Bacterial reagents:
DNA to be mutagenized cloned in the appropriate M13 vector
synthetic oligonucleotide, at least 20 bases long
Luria broth:
 10 g bactotryptone/l,
 5 g yeast extract/l,
 10 g NaCl/l
 thymidine 2 mg/ml
 deoxyadenosine 1 mg/ml
 uridine 0.25 mg/ml

Enzymes
T$_4$ polynucleotide kinase (Boehringer-Mannheim)
DNA polymerase (Klenow), (BRL)
T$_4$ DNA ligase (New England Biolabs)

Radioisotopes
[^{32}P]γ-ATP, 40μCi, S.A. 3000 Ci/mmol

Reagents

20% PEG 6000/2.5 M NaCl
10 × kinase buffer: 500 mMol Tris-Cl pH 7.5, 100 mMol MgCl$_2$
10 mMol rATP
10 mMol each dATP, dCTP, dGTP, dTTP
1 M DTT
3 M sodium acetate, pH 5.5
2 M Tris Cl pH 7.5
1 M MgCl$_2$
5 M NaCl

METHODS

1. Grow M13 template to be mutagenized through 2 cycles of growth in the *dut ung* host. We have followed the procedure outlined by Kunkel (1985).[2] Dilute bacterial overnight 1:100 in 50 ml of Luria broth supplemented with 20 µg/ml thymidine, 100 µg/ml deoxyadenosine and grow until the cells reach a density of 4×10^8 cells/ml (OD$_{600}$ = 0.3). Spin 2000 rpm in RC-2B centrifuge for 10 minutes, wash in unsupplemented Luria broth, and bring up in 50 ml pre-warmed Luria broth supplemented with 0.25 µg/ml uridine. Grow for 5 minutes and infect with 0.5 ml M13 phage stock (approximately 10^{11}–10^{12} phage). Grow overnight, repeat another cycle of growth in the *dut ung* strain, and isolate template in the usual manner for M13 phages. Quantitate the yield of DNA at OD$_{260}$. 1 OD unit = 50 µg/ml DNA.

2. 5′ end label oligonucleotide with polynucleotide kinase: mix 1 µg of oligonucleotide with 2 µl of 10 × kinase buffer, 2 µl of 10 mM rATP, 40 µCi of [^{32}P]γ-ATP, q.s. to 20 µl. Add 2–4 units of T$_4$ polynucleotide kinase. Incubate at 37°C for 60 minutes, then at 65°C for 10 minutes. Add 1/10 volume of 3 M sodium acetate, pH 5.5 and add 2 volumes of ethanol. Quick precipitate at −70°C and spin in the microcentrifuge. Bring the pellet up in 75 µl of water.

3. Solutions:[6]
   Solution A: 0.2 M Tris-Cl, pH 7.5
                0.1 M MgCl$_2$
                0.5 M NaCl
                10 mMol DTT
   Solution B: 0.2 M Tris-Cl, pH 7.5
                0.1 M MgCl$_2$
                0.1 M DTT
   Solution C: 1 µl solution B
                1 µl of 10 mMol dNTP solution
                1 mMol rATP
                3 units DNA polymerase (Klenow)
                1 µl T$_4$ DNA ligase (400 units/µl)
                q.s. to 10 µl

4. Annealing: M13 single stranded DNA 1 µg
   oligonucleotide 8 µl
   solution A 1 µl
   Incubate at 55–58°C for 10 minutes. Keep at room temperature for 10 minutes.
5. Add 10 µl of solution C to the annealed mixture. Take 2 µl out for the gel and freeze this aliquot immediately. Incubate the reaction at room temperature for 5 minutes and then place at 15°C overnight.
6. Take 2 µl out of the reaction mixture for the gel. Make an 0.8% agarose gel with 2 µg/ml ethidium bromide and run both aliquots from the reaction mixture on the gel with both single stranded and double stranded M13 as markers. As the reaction proceeds, the single stranded form should be at least partially converted to the covalently closed circular double stranded form.
7. Transform appropriate dilutions of the reaction mixture into JM101 as described in Chapter 9. Transform a wide range of dilutions to ensure that plaques will be easily isolated. Pick plaques with sterile Pasteur pipets into 0.5 ml of LB, reinfect JM101 with a small amount of the supernatant and replate. After this repurification step, plaques can be picked and the DNA from them analyzed for the desired mutation.

ANALYSIS OF MUTANTS

Localization of the mutant plaques has been accomplished in four ways:
1. biological assay,
2. hybridization,
3. creation/ablation of a restriction enzyme site, and
4. direct DNA sequencing.

The first of these is impractical for most applications since it depends on whether the mutated gene product has an enzymatic activity in *E. coli*. The second of these methods was widely used for analysis of mutants when mutation frequencies were on the order of a few per hundred plaques; of all the techniques available, that using tetramethylammonium chloride is probably the most reliable. The third method can be used if there is a new restriction site created or one ablated by the new mutation. To perform this test, double stranded replicative form DNA is prepared for a number of plaques using conventional techniques for plasmid DNA isolation (see Ch. 2) and then cut by the relevant restriction endonuclease. For these last two methods, once the mutant plaques have been identified the DNA must still be sequenced to confirm the presence of the mutation. For this reason, in our laboratory we prefer the last method as a direct screen for mutants, that of direct DNA sequencing. This usually requires synthesis of an oligonucleotide, just downstream of where the mutation is to be made, for use as a sequencing primer. Single stranded template from a number of plaques is prepared. As a first screen, sequencing reactions are performed for only one of the bases (see Fig. 11.2). The mutants are then subjected to dideoxy sequencing with all four reactions to confirm the mutation. For example, we

154 MOLECULAR GENETICS

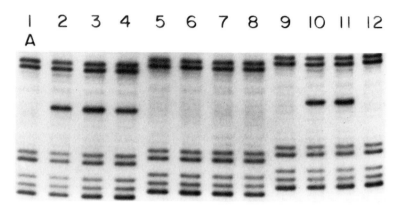

Fig. 11.2 Example of a dideoxy DNA sequencing screen for mutants. A-tracks (DNA sequencing reactions that detect only the A residues) were analyzed simultaneously on DNA from 12 plaques. Mutants occur where there is a loss of adenosine. The mutation frequency here is about 60%.

screen twelve plaques and fully sequence only two. The high mutation frequency (40–80%) achieved by the protocol outlined in this review makes this method the most useful for mutant analysis. After the mutants are identified, double stranded replicative form of M13 is prepared using conventional methods (see Ch. 9), and the mutated fragment is subcloned according to the investigator's needs.

OTHER METHODS

There are some circumstances where mutagenesis directly in a double stranded DNA plasmid is desirable. For example, it is possible that the DNA of interest may be difficult to subclone into the M13 phage, or that it may not be able to be stably cloned into M13. The use of plasmids for oligo-directed site-specific mutagenesis was originally described in 1980[10] and involved generation of a single stranded template by exonuclease III. Subsequently, Schold et al[11] have described a procedure which does not require the generation of a single stranded template from the double stranded plasmid. However, they found that when compared to mutagenesis using M13-generated template, their mutation frequency was much lower. The development of an efficient technique for site-specific mutagenesis directly in plasmid will be an important contribution to the field of genetic engineering.

Other techniques have been described to increase the mutation frequency of oligonucleotide-directed site-specific mutagenesis in M13. These include reports by Norris et al,[12] Carter et al,[13] Kramer et al,[14] and Taylor et al,[15] all of whom have used different procedures to generate mutants at a frequency greater than 40%. Although we do not have any direct experience with these procedures, growth of the template strand in the *dut ung E. coli* achieves the same mutation frequency or better with less manipulation.

PROTOCOL
Alkaline lysis procedure for plasmid preps.

Materials
Chloramphenicol
Lysozyme
10 N NaOH
10% SDS
Cesium chloride
1 l Luria broth (Bactotryptone 10 g, yeast extract 5 g, NaCl 10 g/liter pH to 7.5 with NaOH) in a 4 l flask. Autoclave. Add appropriate antibiotic when cool: ampicillin 35–50 μg/ml, tetracycline 25 μg/ml, kanamycin 50 μg/ml.
Solution I: 20 mM glucose
10 mM Tris, pH 8.0
4 mM EDTA, pH 8.0
Autoclave. Store at 4°C.
Solution III: 3 M potassium acetate
2 M acetic acid
Mix 300 ml 5 M potassium acetate with 58 ml of glacial acetic acid and 142 ml of water. Autoclave. Store at 4°C.
TE — 10 mM Tris, pH 7.5
2 mM EDTA
3 M sodium acetate, pH 5.5.

Method
1. The night before you plan to grow up the 1 l culture, inoculate 10 ml bacterial overnight culture. To make this, inoculate 10 ml Luria broth and appropriate antibiotic with a single colony of bacteria using a sterile loop. Grow at 37° overnight. This will be the seed culture.
2. Inoculate 1 l Luria broth with appropriate antibiotic with 10 ml overnight culture.
3. Grow at 37°C until the $OD_{600} = 0.4$–0.5.
4. Add 1 ml 150 mg/ml chloramphenicol in ethanol.
5. Grow at 37°C 18–20 hours.
6. Harvest cells in sterile 250 ml bottles by spinning at 6K for 5 minutes at 4°C in a Sorvall GSA Rotor. Keep on ice.
7. Resuspend each liter in 50 ml Solution I. Keep on ice.
8. Add 10 ml Solution I containing 60 mg lysozyme.
9. Swirl to mix.
10. Add 20 ml fresh Solution II:
    0.54 N NaOH 20 ml:1 ml 10 N NaOH
    3% SDS   6 ml 10% SDS
    qs to 20 ml in H_2O
11. Gently mix.

12. Incubate 15 minutes on ice.
13. Add 30 ml Solution III.
14. Mix well by swirling.
15. Incubate on ice for greater than 30 minutes.
16. Spin at 10 000 rpm for 20 minutes at 15°C in 250 ml bottle in a Sorvall GSA Rotor.
17. Pour off supernatant into a new 250 ml bottle.
18. Add 65 ml isopropanol.
19. Mix well.
20. Incubate on ice for 5 minutes.
21. Spin out DNA at 10 000 rpm, 10 minutes, 15°C, in a Sorvall GSA Rotor.
22. Decant and drain the pellet well.
23. Add 28 ml TE to resuspend the pellet.
24. Transfer to a 50 ml conical centrifuge tube. Measure volume.
25. Add cesium chloride to DNA solution on a gram per/ml basis. If there are 29 ml of DNA solution, add 29 g cesium chloride.
26. Dissolve well.
27. Add 1 ml 15 mg/ml ethidium bromide. Mix.
28. Transfer to a large (1 × 3 1/2 in) Quick seal centrifuge tube.
29. Balance tubes with 1 g/ml CsCl solution.
30. Centrifuge at 40 000 rpm, 18°C in a VTi 50 rotor for 20 hours.
31. To recover plasmid band: place tube in test tube holder snip off top with scissors draw off band with a 10 ml syringe/20 gauge needle by entering the tube below the band
32. Neutralize contents of tube in bleach, then discard.
33. Extract ethidium bromide from DNA by serial extractions with n-butanol. The DNA solution will be on the bottom.
34. Dialyze CsCl out of DNA in 1 l TE in a 1 liter graduated cylinder for 3 hours.
35. Recover DNA solution, add 1/10 vol 3 M sodium acetate, pH 5.5, and 2 volumes of ethanol to precipitate. Store at −20°C overnight.
36. Spin out DNA 10 000 rpm, 15 minutes in an HB4 rotor.
37. Lyophilize to remove ethanol. Dissolve in 0.5 ml TE. Measure amount by OD_{260}. 1 OD_{260} = 50 $\mu g/ml$ DNA.

REFERENCES

1 Zoller M J, Smith M. Oligonucleotide-directed mutagenesis of DNA fragments cloned into M13 vectors. Methods Enzymol 1983; 100: 468–500.
2 Kunkel T A. Rapid and efficient site-specific mutagenesis without phenotypic selection. Proc Natl Acad Sci USA 1985; 75: 468–492.
3 Hutchison C A III, Phillips S, Edgell M H, Gillam S, Jahnke P, Smith M. Mutagenesis at a specific position in a DNA sequence. J Biol Chem 1978; 253: 6551–6560.
4 Razin A, Hirose K, Itakura K, Riggs A D. Efficient correction of a mutation by use of chemically synthesized DNA. Proc Natl Acad Sci 1978; 75: 4268–4270.

5 Zoller M J, Smith M. Oligonucleotide-directed mutagenesis using M13-derived vectors: an efficient and general procedure for the production of point mutations in any fragment of DNA. Nucleic Acids Res 1982; 10: 6487–6499.
6 Zoller M J, Smith M. Oligonucleotide-directed mutagenesis: a simple method using two oligonucleotide primers and a single-stranded DNA template. DNA 1984; 3: 479–488
7 Fritz H J. The oligonucleotide-directed construction of mutations in recombinant filamentous phages. In: Glover D M, ed. DNA cloning. Vol 1. Oxford: IRL Press, 1985: p 151–163.
8 Gillam S, Smith M. Site-specific mutagenesis using synthetic oligodeoxyribonucleotide primers: I optimum conditions and minimum oligodeoxyribonucleotide length. Gene 1979; 8: 81–97.
9 Kunkel T A, Roberts J D, Zakour R A. Rapid and efficient site-specific mutagenesis without phenotypic selection. Methods Enzymol 154: 367–382.
10 Wallace R B, Johnson P F, Tanaka S, Schold M, Itakura K, Abelson J. Directed deletion of a yeast transfer RNA intervening sequence. Science 1980; 209: 1396–1400.
11 Schold M, Colombero A, Reyes A A, Wallace R B. Oligonucleotide-directed mutagenesis using plasmid DNA templates and two primers. DNA 1984; 3: 469–477.
12 Norris K, Norris F, Christiansen L, Fiil N. Efficient site-directed mutagenesis by simultaneous use of two primers. Nucleic Acids Res 1983; 11: 5103–5112.
13 Carter P, Bedouelle H, Winter G. Improved oligonucleotide-directed site-specific mutagenesis using M13 vectors. Nucleic Acid Res 1985; 13: 4431–4443.
14 Kramer W, Drutsa V, Jansen H W, Kramer B, Pflugfelder M, Fritz H J. The gapped duplex DNA approach to oligonucleotide-directed mutation construction. Nucleic Acids Res 1984; 12: 9441–9455.
15 Taylor J W, Oh J, Eckstein F. The rapid generation of oligonucleotide-directed mutations at high frequency using phosphorothioate-modified DNA. Nucleic Acids Res 1985; 13: 8765–8785.

PART TWO
Oligonucleotide design and use
K. J. Lomax

INTRODUCTION

Oligonucleotides are powerful molecular reagents that have been used with success in a wide variety of experimental techniques. They provide a means of designing specific probes to detect DNA and RNA molecules for which only partial nucleic acid sequence may have been previously determined. If no base sequence is known, a short stretch of amino acid sequence of the relevant protein provides enough information to construct an oligonucleotide that will be complementary to the DNA or RNA of interest. These short chemically synthesized nucleic acid sequences provide many of the capabilities of the cloned gene when used as hybridization probes or primers. They are an invaluable tool in the research laboratory and have become increasingly useful for diagnosis of genetic disorders.

Chemically synthesized oligonucleotides generally range from 8 to 100 bases in length and may be any sequence desired. Longer sequences (greater than 40 bases in length), may be directly synthesized, which becomes increasingly difficult and expensive with increasing length, or may be constructed from a series of shorter sequences ligated together. Both very short or longer oligonucleotides may be used effectively as hybridization probes in DNA or RNA blotting and in screening cDNA libraries for sequences from rare mRNAs. The probes are radioactively labeled and used directly in molecular hybridization assays to detect DNA or RNA molecules having complementary sequences. Development of very stringent hybridization and washing techniques has made it possible to use oligonucleotide probes to detect even single base differences in otherwise complementary regions of base sequence. Unique sequence oligonucleotide probes consist of a single base sequence and are used when a high degree of specificity is desired. In addition, short unique sequence oligonucleotides may be used to prime cDNA synthesis from selected mRNA molecules. These synthetic primers are annealed to mRNA molecules containing a complementary base sequence by Watson-Crick base pairing, and the primer then serves as a starting point for copying of the template sequence by reverse transcriptase or other polymerase enzymes. Mixed sequence oligonucleotide probes consist of 2 to 50 or more probes, usually of the same length, but different nucleotide sequences. They are used together as mixtures in instances where the precise complementary base sequence is not known; for example, when only the amino acid sequence, but not the nucleotide sequence, is known for the region of interest of the gene.

STRATEGIES FOR PROBE AND PRIMER CONSTRUCTION

Strategies for constructing oligonucleotide probes are best worked out in consultation with someone who has had experience with their design and use. However, there are general guidelines which will be helpful to anyone involved in work utilizing these molecular reagents. A successful strategy begins with some knowledge of the nucleotide sequence of interest or the amino acid sequence of the relevant protein. Optimum results will be obtained if the eventual use of the synthetic oligonucleotide is considered at the time of design. The first consideration is whether the oligonucleotide will need to be complementary to mRNA as well as DNA sequences. This will determine the 5' to 3' base sequence orientation. After deciding whether the oligonucleotide will need to be complementary to the sense or antisense strand of DNA, the design of a unique sequence probe is relatively straightforward. Short probes are useful to detect even single base differences under stringent hybridization and washing conditions and have been used to identify carriers of genetic abnormalities caused by a single base change.[1,2] When screening cDNA or genomic libraries or probing genomic DNA, the sequence complementary to a short probe may be found in a number of genes, and the frequency of this occurrence may be estimated based on the length of the oligonucleotide.[3] Longer probes (greater than 20 nucleotides) will generally have more specificity than very short probes.

Mixed sequence oligonucleotide probes are constructed when some amino acid sequence of the relevant protein is known, but no nucleotide sequence data are available. Because of the degeneracy of the genetic code, most amino acids are specified by more than one codon and usually it is not possible to determine exact base sequence from amino acid sequence data. By making probes for all possible choices of nucleotides, you can be certain of having at least one of the mix completely complementary to the DNA or RNA sequence of interest. Although very large mixes of probes have been used successfully (greater than 100 in a mix), the sensitivity of these mixed probes may be optimized by holding the numbers of individual probes in the mix to a minimum.[4] For that reason, these probes are kept relatively short and usually correspond to a 5 to 8 amino acid sequence. Careful selection of the amino acid sequence to use as a template is essential in preventing the number of probes in the mix from becoming unwieldy. Amino acid sequences with minimal codon degeneracy will allow the most accuracy in predicting nucleotide sequence (Table 11.2). Tryptophan and methionine are therefore the most desirable to use, because they each have only a single codon and their corresponding nucleotide sequence can be exactly predicted. Nine other amino acids (PHE,TYR,HIS,GLN, ASN,LYS,ASP,GLU,CYS) have only two codons with the ambiguity residing in the third position. For six of these, the third nucleotide is either U or C.[5] If the nucleotide sequence of interest is a member of a family of protein genes, some of whose nucleotide sequences are known, or if a single gene from several different species has been sequenced, it may be possible to reduce ambiguity by selecting nucleotides based on codon usage within these genes.[6,7]

Table 11.2 The genetic code.

Amino acid	Codon
Alanine	GCU
	GCC
	GCA
	GCG
Arginine	CGU
	CGC
	CGA
	CGG
	AGA
	AGG
Asparagine	AAU
	AAC
Aspartate	GAU
	GAC
Cysteine	UGU
	UGC
Glutamate	GAA
	GAG
Glutamine	CAA
	CAG
Glycine	GGU
	GGC
	GGA
	GGG
Histidine	CAU
	CAC
Isoleucine	AUU
	AUC
	AUA
Leucine	CUU
	CUC
	CUA
	CUG
	UUA
	UUG
Lysine	AAA
	AAG
Methionine	AUG
Phenylalanine	UUU
	UUC
Proline	CCU
	CCC
	CCA
	CCG
Serine	UCU
	UCC
	UCA
	UCG
	AGU
	AGC

Table 11.2 Contd.

Amino acid	Codon
Threonine	ACU
	ACC
	ACA
	ACG
Tryptophan	UGG
Tyrosine	UAU
	UAC
Valine	GUU
	GUC
	GUA
	GUG

In addition to their use as hybridization probes, synthetic oligonucleotides are also excellent primers for cDNA synthesis, particularly when one desires to transcribe a rare mRNA species within a cellular mRNA sample. They offer a simple means of selecting a rare mRNA from a heterogeneous population and enriching cDNA synthesis from that template. After annealing to the mRNA, the primer serves as a starting point for cDNA synthesis using reverse transcriptase or other polymerase enzymes. These cDNA products may be used directly as hybridization probes, which, because of their greater length and higher specific activity, will provide greater specificity and sensitivity than is possible with synthetic oligonucleotides. Alternatively, after second strand synthesis, these cDNAs may be cloned into vector phage or plasmid to create a cDNA library.

When designing a synthetic oligonucleotide to be used as a primer, there are a number of things to consider. Stability of the interaction between the mRNA and primer is dependent on the length of the oligonucleotide, percentage of G-C pairings, numbers of potential mismatches and secondary structure of the mRNA. Efficient primers are at least 8–10 nucleotides long with 50–55% G-C content to maximize the strength of the Watson-Crick base pairing with the mRNA. Although some mismatches between primer and mRNA may be allowed (dG-U, dT-G), a 3' terminal dG-U will not allow initiation of reverse transcription.[5] Under the usual conditions for annealing, some mismatches will be tolerated and even unique sequence oligonucleotide primers will produce some cDNA sequences transcribed from regions other than those of interest. Mixed sequence oligonucleotide primers would be expected to amplify this problem. As is true for very short oligonucleotides when used as hybridization probes, these short sequences when used as primers can be expected to hybridize to more than one sequence complementary to themselves. If the mRNA is involved in stable intrastrand interactions, it may be difficult for the primer to hybridize. Adequate denaturation of the mRNA template will improve the ability of the primer to hybridize efficiently, however there will still be some variability in efficiency of hybridization from primer to primer. To allow the reverse transcriptase to transcribe the desired region of the mRNA, the oligonucleotide primer should be chosen to be complementary to a position 3' to the targeted region within the mRNA molecule.

GENERAL CONSIDERATIONS FOR USE OF OLIGONUCLEOTIDES

Standard protocols for Southern blotting, Northern blotting, screening of cDNA and genomic libraries and primer extension have been modified to adapt to the special requirements of synthetic oligonucleotide probes and primers. These modifications attempt to provide sufficiently stringent hybridization and washing conditions to allow perfect match hybridization, while preventing hybridization to partly complementary, but not identical, sequences. Binding of oligonucleotides is dependent on their length and guanosine-cytosine (G-C) and adenosine-thymine (A-T) content. The oligonucleotide dissociation temperature (T_d) can be determined empirically and provides a point of reference for determining hybridization and wash temperatures for unique sequence probes.[8,9] The equation for dissociation of an oligonucleotide probe from RNA or DNA is as follows;

$$T_d = (\# \text{ of G-C base pairs}) 4°C \times (\# \text{ of A-T base pairs}) 2°C$$

Blots are usually hybridized at 10–20°C below the T_d and washed at 5–10°C below T_d. After washing, the filters may be scanned while wet with a Geiger counter and, if background seems too high, washed at a temperature closer to the T_d. When mixed sequence probes having the same length but variable base composition are used, suitably stringent conditions become increasingly difficult to achieve for all members of the pool, and a number of false positives may occur. This is especially true when probing genomic DNA or screening complex libraries for genes of low abundance. This problem can be minimized by an alternative protocol utilizing tetramethylammonium chloride (TMAC). TMAC, by binding selectively to A-T pairs, displaces the dissociation equilibrium and eliminates dependence of T_d on G-C content. In the presence of 3 M TMAC, T_d is solely dependent on probe length.[3] This method will allow only perfect match hybridization along the probe length chosen and will minimize false positives when using mixed sequence probes.

There are several other considerations when using oligonucleotide probes which may require changes in methods, in addition to modifying hybridization and wash temperature. Both nitrocellulose and nylon filter membranes may be used for blotting. Nylon transfer media have gained in popularity because they can be washed free of probe after use and reprobed repeatedly. However, washing conditions may need to be more stringent than is the case with nitrocellulose filters in order to achieve an acceptable signal-to-noise ratio. Addition of Denhardt's solution and heterologous DNA to the hybridization buffer will decrease binding of the probe, and these constituents are omitted from that step when using nylon filters. Radioactive labeling of small oligonucleotide probes by

```
                    A
        5'-AATTTATCTGGGCAATC-3'
              G  C   A  G
                 G
```
17-mer homologous to Lf

Fig. 11.3 This mixed sequence 17-mer probe was constructed to represent all possible codon choices for a short amino acid sequence (N'–ASP CYS PRO ASP LYS PHE–C') of lactoferrin. One arbitrary nucleotide choice was made at position 3 to reduce the number of individual oligonucleotides in this mixed probe.

kinasing is done according to standard protocols.[10] Specific activity should approach 1×10^8 cpm/μg. To remove unincorporated isotope, probes may be precipitated several times with ammonium acetate, or passed over G-25 fine Sephadex (Pharmacia) columns before use. Probes should be boiled for several minutes and quick-chilled on ice before adding to the hybridization buffer at a final concentration of 1×10^6 cpm/ml. For mixed sequence probes, it may be necessary to take into account the number of individual probes in the pool and increase the total amount of probe used in the hybridization to adequately represent all members of the pool.[3]

GENERAL PROTOCOLS UTILIZING OLIGONUCLEOTIDE PROBES AND PRIMERS

Southern blotting

Southern blotting techniques may be used to probe DNA restriction fragments from a variety of sources, and oligonucleotides have been used widely as probes in this type of experiment. More recently, highly stringent conditions for hybridization and washing have allowed use of very small oligonucleotides to probe restriction enzyme digested genomic DNA and can detect as little as a single base change. These techniques have provided a powerful new tool for prenatal diagnosis of several genetic diseases.[1,2] Preparation of the gel and DNA transfer to a filter are done by standard methods (cf Ch. 2).[10] The procedure that follows is adapted from Wirth for use with nylon filters and is a general method for Southern blotting.[9]

Pre-hybridization
Place the blot in a plastic sealable bag with a generous amount of pre-hybridization buffer (0.3 ml/cm² blot). Shake 3 hours to overnight at the hybridization temperature (see below). Remove the filter from the bag and rinse it off with a small amount of hybridization solution before placing it in a clean bag. This should remove any residual pre-hybridization solution containing Denhardt's solution and heterologous DNA which will inhibit hybridization of the probe.

Pre-hybridization buffer
6 × SSC
1% SDS
10 × Denhardt's solution
Heat to 55°C and filter. Add 20 µ/ml *E. coli* or yeast tRNA and 50 µg/ml salmon sperm (SS) DNA.

Hybridization
Add just enough hybridization solution to cover the blot (usually 0.1 ml/cm² blot). Probe concentration should be 1×10^6 cpm/ml buffer. Hybridize in a shaking water bath at 10–20°C below calculated T_d for the probe for 12 hours.
Hybridization buffer
6 × SSC
1% SDS
Heat to 55°C and filter before use.

Washing conditions
Remove the filter from the hybridization bag and place in 100–200 ml (20 ml/cm² blot) of wash buffer. Shake at room temperature for 15–20 minutes. Pour off the buffer and add fresh buffer pre-heated to 5°C below the T_d of the probe. Wash for 15–20 minutes, replace the buffer with fresh buffer and wash an additional 15–20 minutes. If background seems high upon scanning the filter with a Geiger counter, a final 10 minute wash in 1 × SSC and 1% SDS at the T_d of the probe may be necessary. Autoradiograph the filter with an intensifying screen at −70°C for 72–96 hours. A shorter or longer exposure may be necessary.
Wash buffer
6 × SSC
1% SDS

Northern blotting
Oligonucleotide probes, both mixed or unique sequence, may be used to rapidly identify rare mRNAs using Northern blotting techniques. This circumvents some of the difficulties inherent in earlier techniques for studying mRNAs utilizing immunoradiometric assays requiring antibodies to an mRNA protein product which may undergo significant post-translational modification or may be present in such small amounts that it is impossible to obtain an antibody. It also provides a means of studying RNAs without prior knowledge of the nucleotide sequence if at least some of the relevant amino acid sequence is known. Total or poly-A⁺ RNA is run on a formaldehyde gel according to standard methods (see Ch. 3).[10] RNA is transferred to the filter overnight via capillary action in 20 × SSC without any pre-treatment of the gel. The protocol that follows is for use with nitrocellulose filters.[13]

Pre-hybridization
Place the blot in a sealable plastic bag with a generous amount of pre-

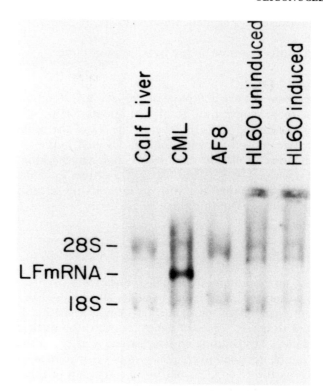

Fig. 11.4 A Northern blot was hybridized to the mixed sequence 17-mer probe described in Figure 11.3 Lane 1 — calf liver 18 and 28S ribosomal RNA markers. Lane 2 — poly-A$^+$ RNA isolated from the peripheral leukocytes of a patient with chronic myelogenous leukemia. Lane 3 — poly-A$^+$ RNA from a Syrian hamster kidney fibroblast cell line. Lane 4 — poly-A$^+$ RNA from uninduced HL-60 cells. Lane 5 — poly-A$^+$ RNA from HL-60 cells induced for 3 days with DMSO. A band is seen in Lane 2 corresponding to the expected size of the lactoferrin mRNA transcript. Other faint bands represent non-specific hybridization.

hybridization buffer (0.3 ml/cm^2 blot). Incubate in a shaking water bath for a minimum of one hour at the hybridization temperature (see below).

Pre-hybridization buffer:
4 × SSC
5 × Denhardt's solution
0.2 mM EDTA
0.1% SDS
Heat to 55°C and filter before use.

Hybridization
Squeeze out pre-hybridization buffer and add just enough hybridization buffer to cover the blot (0.1 ml/cm^2 blot). Probe concentration should be 1 ×

10^6 cpm/ml. Hybridize in a shaking water bath for 12 hours at 10–20°C below T_d of the probe.

Hybridization buffer — same as pre-hybridization buffer.

Washing conditions
Remove the filter from the bag and place in 100–200 ml (20 ml/cm² blot) of wash buffer. Shake at room temperature for 15–20 minutes. Pour off the buffer and add fresh buffer pre-heated to 5–10°C below the T_d of the probe. Wash for 10 minutes and replace the buffer with fresh pre-heated buffer and wash for an additional 10 minutes. If background seems high on checking with a Geiger counter, a final wash in 1 × SSC and 0.1% SDS at the T_d of the probe may be necessary. Autoradiograph the filter with an intensifying screen at −70°C overnight. A shorter or longer exposure may be necessary.

Wash buffer
2 × SSC
0.1% SDS

Screening cDNA and genomic DNA phage libraries

Synthetic oligonucleotides have proved to be an efficient and versatile means of identifying recombinant DNA molecules containing a specific DNA sequence. They can be used to detect clones of genes for which no nucleic acid sequence was previously known. Depending on what amino acid or nucleic acid sequence is known, they can be designed to recognize clones that contain a specific portion of the gene such as the 5' end, 3' end or an important coding region. Plating of cDNA libraries and handling of filters has been described elsewhere (Chs. 5–7).[10] The following protocol describes hybridization and washing of filters and was adapted from Woods.[11]

Pre-washing
Pre-washing the filters removes bits of agarose, bacteria, etc. to leave a clean filter for hybridization and reduce background. An hour or two of pre-washing in a large volume of buffer (500 ml/40 filters) at 65°C may suffice for filters that were relatively clean to start with. After pre-washing, filters are immediately pre-hybridized or air dried and stored at 4°C.

Pre-wash buffer
3 × SSC
0.1% SDS

Pre-hybridization
Filters may be stacked in sealable plastic bags or placed in a plastic box with 200 ml buffer/40 filters. Shake in a water bath at the hybridization temperature (see below) for 2 hours to overnight.

Pre-hybridization buffer
6 × SSC
1 × Denhardt's

0.5% SDS
100 μg/ml salmon sperm DNA
0.05% sodium pyrophosphate

Hybridization
Filters are divided into separate plastic bags with ten stacked filters in each bag. 1 ml/filter of hybridization buffer containing probe at a concentration of 1×10^6 cpm/ml is added to each bag. Bags are placed in a shaking water bath at 10–20°C below T_d of the probe and hybridized 12–24 hours.

Hybridization buffer
6 × SSC
1 × Denhardt's
20 μg/ml yeast tRNA
0.05% sodium pyrophosphate

Washing conditions
Remove filters from bags and rinse one at a time in wash buffer at room temperature. Place them in a large plastic box containing approximately 20 ml/filter pre-heated wash buffer and shake for one hour at the hybridization temperature. Transfer to another box containing the same volume of wash buffer pre-heated to 5°C below T_d of the probe and shake at that temperature for 20 minutes. Repeat that step for two additional washes. Air dry filters and autoradiograph at −70°C with an intensifying screen overnight. Shorter or longer exposure time may be necessary. Positive signals should be clearly apparent over background.

Wash buffer
6 × SSC
0.05% sodium pyrophosphate

Genomic screening
Either the method described above or the TMAC protocol may be used for screening genomic libraries.[3] Because of their complexity, screening these libraries with oligonucleotide probes may result in a number of positive signals resulting from the probability of perfect matches occurring not only in the gene of interest, but randomly throughout the genome. A means of calculating expected numbers of sequence matches based on complexity of the genome and probe length may be found in Wood et al.[3]

PRIMER EXTENSION

Primer extension utilizing short synthetic oligonucleotides as primers allows you to select out rare mRNAs from a pool of RNAs and enrich cDNA synthesis from those templates. The amount of cDNA synthesized is dependent on the amount of mRNA available to use as template. For very low abundance mRNAs, isolation of amounts sufficient for primer extension may be an arduous task.

However, this technique has been used with success with mRNAs representing as little as 0.005% of the total RNA extracted from tissue.[12] The cDNA products may be put to a variety of uses, and primer extension protocols reflect the intended use of those products. If the cDNA products are to be sequenced, the oligonucleotide primers are kinased prior to annealing to mRNA. The primer extension products can be run on a 12% polyacrylamide/urea gel and promising bands can be eluted and directly sequenced according to the method of Maxam & Gilbert. If the cDNA products are to be used as probes, radioactive nucleotides are used in the cDNA synthesis to produce very high specific activity probes.[14] Oligonucleotide primers may also be used to prime first strand cDNA synthesis to enrich for rare mRNAs and avoid the more laborious task of size-selecting or hybridization-selecting the mRNA itself.

REFERENCES

1 Conner B J, Reyes A A, Morin C, Itakura K, Teplitz R L, Wallace R B. Detection of sickle cell Beta^s-globin allele by hybridization with synthetic oligonucleotides. Proc Natl Acad Sci USA 1983; 80: 278–282.
2 Saiki R K, Scharf S, Faloona F, Mullis K B, Horn G T, Erlich H A, Arnheim N. Enzymatic amplification of β globin genomic sequences and restriction site analysis for diagnosis of sickle cell anemia. Science 1985; 230: 1350–1354.
3 Wood W I, Gitschier J, Lasky L A, Lawn R M. Base composition-independent hybridization in tetramethylammonium chloride: A method for oligonucleotide screening of highly complex gene libraries. Proc Natl Acad Sci USA 1985; 82: 1585–1588.
4 Whitehead A S, Goldberger G, Woods D E, Markham A F, Colten H R. Use of a cDNA clone for the fourth component of human complement (C4) for analysis of a genetic deficiency of C4 in guinea pig. Proc Natl Acad Sci USA 1983; 80: 5387–5391.
5 Agarwal K L, Brunstedt J, Noyes B E. A general method for detection and characterization of an mRNA using an oligonucleotide probe. J Biol Chem 1981; 256(2): 1023–1028.
6 Ullrich A, Shine J, Chirgwin J, Pictet R, Tischer E, Rutter W J, Goodman H M. Rat insulin genes: Construction of plasmids containing the coding sequences. Science 1977; 196: 1313–1319.
7 Efstratiadis A, Kafatos F C, Maniatis T. The primary structure of rabbit β globin mRNA as determined from cloned DNA. Cell 1977; 10: 571–585.
8 Suggs S V, Hirose T, Miyake T, Kawashime E H, Johnson M J, Itakura K, Wallace R B. In: Brown D D, ed. ICN-UCLA Symposium developmental biology using purified genes. 23: 683–693. New York: Academic Press, 1981.
9 Wirth R. Southern blot hybridization with oligonucleotide probes on Schleicher and Schuell Nytran membranes. Sequences 1986; 390: 1– 4.
10 Maniatis T, Fritsch E F, Sambrook J. Molecular cloning — a laboratory manual. New York: Cold Spring Harbor, 1982.
11 Woods D E. Oligonucleotide screening of cDNA libraries. Focus 1984; 6(3): 1–2.
12 Noyes B E, Mevarech M, Stein R, Agarwal K L. Detection and partial sequence analysis of gastrin mRNA by using an oligonucleotide probe. Proc Natl Acad Sci USA 1979; 76(4): 1770–1774.
13 Stetler D, Das H, Nunberg J H et al. Isolation of a cDNA clone for the human HLA-DR antigen α chain by using a synthetic oligonucleotide as a hybridization probe. Proc Natl Acad Sci USA 1982; 79: 5966–5970.
14 Tsonis P A. cDNA probes from oligonucleotides. B M Biochemica 1986; 3(1): 10.

12
Use of *in vitro* translation techniques to study gene expression

S. A. Liebhaber

INTRODUCTION

In vitro experimental systems which translate mRNA into protein have been available for over 25 years. These systems were initially designed to study the biochemistry of translation and their use was limited to a small number of laboratories. With the advent of recombinant DNA research, these systems have become important tools for genetic analysis as well. In parallel with the increased applications of *in vitro* translation systems, technical modifications have been made which increase their general usefulness and make them available to a wider spectrum of investigators. This chapter summarizes the basic characteristics of *in vitro* translation systems, offers details on the most widely used system, rabbit reticulocyte lysate, and discusses the major applications of this system to the study of gene expression.

THE FEATURES OF AN *IN VITRO* TRANSLATION SYSTEM

In general terms, the utility of an *in vitro* translation system can be defined by the following parameters:

1. The system must be able to translate mRNA into protein with an efficiency adequate for subsequent protein analysis. Because these systems do not produce enough protein for analysis by staining or UV absorption, the *in vitro* translation products are usually radiolabeled and analyzed by gel electrophoresis with subsequent autoradiography.

2. The translational activity of the system should be dependent upon addition of exogenous mRNA. The elimination of endogenous mRNA (for example rabbit globin mRNA in rabbit reticulocyte lysate), constitutes the single most important advance in the preparation of *in vitro* translation systems. This modification, reported in 1976,[1] makes more of the translational machinery available for the mRNA of interest.

3. In order for these systems to be generally applied in laboratories with a wide range of technical expertise, they should be relatively easy to use. This is the case with the two most frequently used systems; rabbit reticulocyte lysate (RRL) and wheat germ extract (WGE). Both of these systems can be prepared using relatively simple and rapid protocols (for detailed protocols on RRL and WGE see

references [2] and [3], respectively). Furthermore, both of these systems are available commercially as ready-to-use kits (from Amersham, Bethesda Research Laboratories, and New England Nuclear, to name a few of a rapidly expanding list).

THE *IN VITRO* TRANSLATION SYSTEMS AVAILABLE

In vitro translation systems can be prepared from the cytosolic homogenate of just about any cell of interest. For example, a general protocol is available for preparing *in vitro* translation systems from a variety of tissue culture lines.[4] Translationally active preparations isolated in this way may be used in the study of cell- or tissue-specific translational control mechanisms. However, for the general purpose of using an *in vitro* translation system as a tool for genetic investigation, these specialized systems are not necessary. In general, one of three commonly used systems is usually sufficient; RRL, WGE, and frog oocyte. The RRL system is by far the most generally useful. However, before dealing with the details of RRL, a few points should be made regarding the other two systems.

Wheat germ extract
The main advantage of WGE is that it is the least expensive and simplest to prepare of the three systems mentioned. Wheat germ can be obtained from General Mills free of charge (or bought in the grocery, although such commercial lots are more variable in their translational activity) and prepared with only the use of a centrifuge and a single gel filtration column. Alternatively, it can be purchased commercially either with or without micrococcal nuclease treatment (see below). While usually not the system of choice for reasons detailed below, there are certain situations for which WGE offers specific advantages. For example, the RRL contains a large amount of rabbit globin protein which can interfere with the detection of proteins in the 14 kd range. This problem can be circumvented by switching to WGE. This system can also be used when trying to synthesize and analyze proteins in the size range of 45 kd; $[^{35}S]$methionine labeling in the RRL is often associated with an artifact band in this region which results from the direct transfer of the ^{35}S label to a pre-existing lysate protein. Other alternatives, such as using Tricine-stabilized $[^{35}S]$methionine (NEN) or switching to $[^{35}S]$cysteine, can also be used to circumvent this problem. The main reasons that WGE is not more widely used are that its translation efficiency is lower than that of the other two systems mentioned, and that it is much less successful in synthesizing full-length proteins from long mRNAs.

Frog oocyte microinjection
The frog oocyte is strictly speaking an *in vivo*, not *in vitro*, translation system. However this system should be considered briefly in the context of *in vitro* systems as it can offer specific advantages in the translation of exogenous mRNAs. This system, first introduced in 1971,[5] offers the highest translational efficiency of any of the available systems. mRNAs, once microinjected into the cytoplasm of the frog oocyte, continue to be translated for a period of days as

compared to a maximum of 30–60 minutes in most *in vitro* systems. Also, the proteins synthesized in the oocyte are fully modified; when appropriate, signal peptides are cleaved, and proteins are glycosylated, phosphorylated and secreted into the media. This system would therefore be an appropriate choice for synthesizing a secretory protein from a rare mRNA for bioassay: in this capacity this system has been used for the identification of interferon mRNA and the first interferon cDNA clones. Disadvantages of the oocyte system include the technical difficulty and expense of maintaining frogs, isolating eggs, microinjecting mRNA and processing the labeled egg cytosol for gel analysis. Details of this system have been recently reviewed.[6]

THE RABBIT RETICULOCYTE LYSATE TRANSLATION SYSTEM

In general, RRL offers the best mix of efficiency, ease of use and versatility for most experimental needs. In this section, the basic components and special modifications of this system are described.

Components

The RRL is a hypotonic lysate of packed red blood cells which are isolated from anemic rabbits and which contain approximately 90% reticulocytes. As such, the lysate itself contains the basic constituents necessary for translation. The reticulocyte lysate (prior to nuclease treatment) will translate its endogenous rabbit globin mRNA with only a few additions.

In order to support translation, the system must be supplemented with *hemin*. If hemin is not added to the system, translation will decrease markedly within 5 minutes. This results from an inactivation (phosphorylation) of a critical initiation factor (eIF2A) (for review of this and other biochemical aspects of the eukaryotic translation system, see reference [7]). The presence of hemin blocks this reaction. Since this inactivation is irreversible, hemin is added to the RRL immediately after its preparation.

An *energy generating system* is necessary as translation initiation and elongation depend upon the availability of the high energy phosphates in the form of GTP and ATP. The pool of high energy phosphates can be maintained during translation either by addition of creatine phosphate and creatine phosphokinase or by the direct addition of GTP and ATP.

Although the reticulocyte lysate already contains *amino acids*, all 20 are added to the reaction mix to ensure that the system does not become starved for substrate. One or more of the amino acids is usually added in a radiolabeled form so that the synthesized proteins can be more easily analyzed.

The most frequently used *radioactive tracers* are [^{35}S]methionine and [^{3}H]leucine. If the protein of interest is particularly rich in a specific amino acid, a labeled form of this amino acid can be added to the system to increase the specific labeling efficiency. For example, labeled proline will result in highly specific labeling of collagen, and labeled cysteine will result in highly specific labeling of metallothienine. In choosing between [^{35}S]methionine and [^{3}H]leucine for general labeling purposes, several parameters should be kept in mind:

1. [^{35}S]methionine can result in a 45 kd background band on gel analysis (see above),
2. ^{35}S does not require fluorography (although this increases the signal severalfold) while ^{3}H must be fluorographed to visualize bands,
3. ^{35}S has a shorter half life (87 days) than ^{3}H (12 years), although this is usually not a concern.

Other constituents added to the *in vitro* translation reaction include buffers to maintain the pH in the optimal range (example: HEPES, pH 7.2) and a reducing agent to stabilize translational factors.

In vitro translations are critically affected by the levels of K^+ and Mg^{2+} present in the translation mix. Each batch of RRL must be individually adjusted and optimized for the concentration of both of these cations in order to achieve maximal translational efficiency for any given mRNA preparation. In addition, in a given system, each individual mRNA species may have its own specific optimal K^+ and Mg^{2+} requirements, and this should be determined when adjusting the conditions for the maximal synthesis of a particular protein.

Special modifications of the RRL

1. The major modification of the RRL is the treatment of the lysate with the enzyme micrococcal nuclease. In the initial IVT systems established in 1963,[8] translation was primarily directed by endogenous mRNA. The activity of the endogenous mRNA often swamped out detection of translation from added (exogenous) mRNA and therefore precluded their translational analysis. Several attempts to decrease endogenous translational activity led to the successful development of an *in vitro* translation system fully dependent for translational activity upon the addition of exogenous mRNA. This was achieved by digesting the endogenous mRNA with micrococcal nuclease (see below), and then inactivating the nuclease before adding exogenous mRNA to the system.[1] This procedure is based upon two specific properties of micrococcal nuclease:

   a. The nuclease appears to be fairly specific for mRNA. The tRNA and rRNA present in the lysate appear to maintain their activity during the nuclease digestion.

   b. The micrococcal nuclease is completely Ca^{2+} dependent. Therefore, the enzyme is added to the lysate along with Ca^{2+}, and, after the desired period of digestion, it is inactivated by adding the Ca^{2+}-chelator, EGTA. The divalent cation needs of the translational system are met by the addition of Mg^{2+} ions, which do not complex with EGTA.

2. A second optional modification of the RRL system is the addition of microsomal membranes capable of processing newly translated proteins.[9] The RRL does not itself contain the rough endoplasmic reticulum (RER) necessary for signal peptide cleavage (globin is a cytosolic protein synthesized on free polysomes). In order to add this capacity to the system, microsomal membranes isolated from dog pancreas (the pancreas is a tissue very rich in RER) are added to the lysate. Translation is then carried out as usual. This modified system can

be used to analyze the synthesis of proteins which contain signal peptides; proteins which are secreted from the cell, or which are located in a cellular vacuolar space such as lysosome, Golgi, endoplasmic reticulum, secretory granules, or which are integral to the cell membrane. The dog pancreas membranes can be purchased commercially or prepared according to published procedures.[10] The membranes will result in the removal of the signal sequence from most, but not all, of the primary protein translation products so that some processed protein and a residual amount of preprotein will be present on final analysis. The analysis of a protein translated in the absence and in the presence of microsomal membranes can be compared to provide direct evidence for the presence of a signal sequence on the initial translation product. In some cases the dog pancreas microsomal membranes may also add sugar residues (mannose-rich sugar complexes) as occurs in the RER *in vivo*. These additions may alter the apparent molecular weight of the processed protein and can usually be removed with the enzyme, Endo H.

ISOLATION OF MESSENGER RNA (see Ch. 3)

The success of an *in vitro* translation system depends to a large extent upon the quality of the exogenous mRNA added to the system. Maximal efforts must be taken to avoid ribonuclease contamination at all steps of RNA preparation and handling. In addition, the presence of trace contaminants from the RNA isolation procedure such as phenol, ethanol, or chloride ions must be minimized in order not to poison the translation system. Often the best RNA isolation procedure must be arrived at empirically. We have had good success translating total RNA extracted from reticulocytes by phenol extraction of acid precipitated polysomes[11] and our best results in the translation of mRNA from tissue and tissue culture have been with RNA isolated by the guanidine hydrochloride method.[12] While total cellular RNA can be translated in the *in vitro* systems, a higher translational efficiency can often be achieved by using purified mRNA isolated by affinity chromatography on oligo-dT cellulose or poly-U sepharose.[12] New technical approaches to this affinity purification are available using oligo-dT linked to a paper support for isolation of mRNA from very small RNA samples.[13]

ANALYSIS OF TRANSLATION PRODUCTS

The analysis of an *in vitro* translation reaction depends upon identification of specific protein products. Because the *in vitro* system will not synthesize sufficient protein to be visualized by staining procedures or UV absorption, detection of the *in vitro* synthesized protein almost always depends upon the measurement of the radiolabeled translation product (the main exception being the bioassay of proteins secreted from the frog oocyte mentioned above). In most cases, the labeled proteins are size fractionated by SDS-polyacrylamide gel electrophoresis (SDS-PAGE) followed by autoradiography. For specific identification of a protein, it is often necessary to demonstrate immunoreactivity to a monospecific antisera. This is usually done by immunoprecipitation followed by SDS-PAGE analysis. In some cases, analysis of the translation reaction by SDS-PAGE is not the optimal method and proteins can be resolved more effectively by other methods such as column chromatography (including HPLC or FPLC), affinity chromatography, or two-dimensional gel electrophoresis. For example, all of the globin proteins are approximately the same size and therefore do not separate well on the SDS-PAGE system. It is therefore necessary to separate them using techniques such as Triton-Acid-Urea gel electrophoresis (separation by hydrophobic interactions), isoelectric focusing (separation by differences in isoelectric points) or cation exchange chromatography (separation by differences in net charge).

USE OF IVT AS A TOOL IN GENETIC ANALYSIS

In vitro translation systems have been used for two main purposes. First, they have been studied *per se* in order to define the biochemistry of translation and translational controls over gene expression. Such studies, although quite productive and of increasing interest, are not in the scope of the present discussion. Second, they have been used as a tool for genetic analysis. In this second case, the most frequent task has been the identification of the protein products encoded by cloned genes or gene fragments. The approach used in this second task is usually based upon the ability of a cloned gene (DNA) to hybridize specifically and with high affinity to its homologous mRNA. This hybridization is based upon Watson-Crick base pairing between the mRNA and the complementary DNA strand. The formation of the hybrids can then be used either to specifically inhibit the translation of the hybridized mRNA or to selectively purify this mRNA for subsequent translation. These two approaches, diagrammed in Fig. 12.1, are known as hybrid-arrested translation (HART) and hybrid-selected translation (HST) respectively.

Hybrid-arrested translation
This was the first of the two approaches to be developed.[14,15] Although still frequently used, it is clearly the weaker of the two techniques because its interpretation is based upon a negative result. In order to identify the protein encoded by a cloned fragment of DNA (a cDNA or a genomic clone) the DNA is hybridized to a sample of mRNA known to contain the homologous mRNA. After hybridization, the mRNA-DNA mixture is purified and translated *in vitro*. The labeled translation products are run on a gel next to an *in vitro* translation of an aliquot of the mRNA-cDNA mixture which has been heat-melted to disrupt hybrids, and the two lanes are compared on the autoradiograph (see Fig. 12.1). The absence of a band in the HART lane implies specific arrest of a specific mRNA species by the cloned DNA. Interpretation of the experimental result is based upon two assumptions:

1. if the mRNA is hybridized to the cloned DNA its translation will be arrested, and conversely,

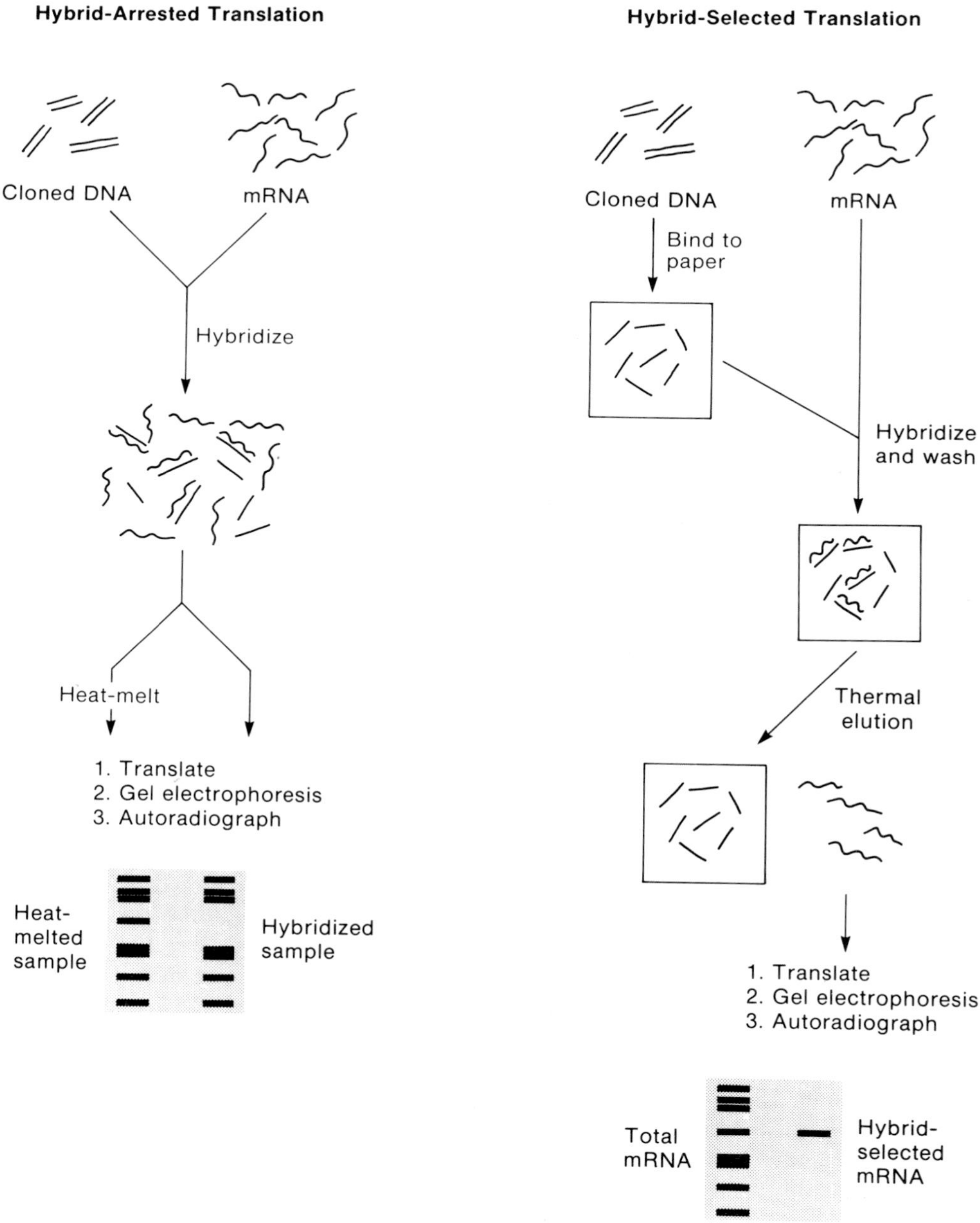

IN VITRO TRANSLATION AND GENE EXPRESSION 175

Hybrid-Arrested Translation

Hybrid-Selected Translation

Fig. 12.1 Hybrid-arrested and hybrid-selected translation. See text for details.

2. the loss of *in vitro* synthesis of a protein by the HART procedure results from specific hybridization to the cloned DNA.

These assumptions are often incorrect and lead to false positive and false negative results:

1. False positive results. Certain mRNAs are translated poorly after incubation under hybridization conditions. This may result from the folding and internal base pairing of the mRNA (secondary structure formation) totally unrelated to DNA hybridization. In this way, loss of translation subsequent to the hybridization reaction can simulate hybrid-arrest and create a false positive result. For this reason, mRNA subjected to a mock hybridization (i.e. a hybridization containing no DNA) must be translated and analyzed as a parallel control.

2. False negative results. Not all DNA fragments can arrest the translation of their complementary mRNA.[16] The only DNA fragments which will reliably hybrid-arrest translation are those which cover the 5' non-translated region of the mRNA. DNA hybridized to the coding region or the 3' non-translated region may have little or no effect upon translation. Unfortunately the 5' non-coding region is the region least likely to be isolated by most cDNA cloning procedures and is a region which will constitute a very small fraction of a genomic clone.

For all the reasons enumerated above, the HART system should usually be avoided except in particularly well defined situations.

Hybrid-selected translation

For the reasons stated above, the HART technique has given way to the more definitive technique of hybrid-selected translation (HST) for most applications. As in the case of HART, HST depends upon specific hybridization of a cloned DNA fragment to its complementary mRNA species (see Fig. 12.1). In contrast to the HART system, the HST hybridization is carried out with the DNA fixed to a solid support, usually nitrocellulose paper. After the hybridization reaction is complete, non-hybridized mRNA is washed off the paper, and the remaining specifically hybridized mRNA is then removed by thermal elution. This hybrid-selected mRNA is then translated *in vitro* and the radiolabeled protein product is analyzed. As in the case of HART, a prerequisite of the procedure is the availability of an RNA sample containing the mRNA species of interest. The HST technique results in the positive identification of the protein encoded by any cloned fragment of DNA. Once a cDNA clone has been isolated and tentatively identified, the hybrid selection system can be used for confirmation of the identification by selecting out the specific encoded mRNA and then translating it into protein. If the protein migrates at the size predicted, its identity can be further confirmed by demonstrating specific immunoprecipitation with a monospecific antibody. If it is desirable to determine whether or not the protein has a signal peptide, the hybrid-selected mRNA can be translated in the presence and the absence of dog pancreas membranes (see above). In contrast to the HART approach, a cDNA or genomic DNA fragment complementary to any region of the mRNA will give a positive result with HST. False positive results with HST can be minimized by carrying out the hybridizations at a high stringency and by confirming positive results with immunoprecipitation. The difficulties in using HST lie in the multiple steps involved, each of which must be done with care to avoid RNA degradation.

The hybrid-selection approach can be modified in a number of ways. For example, a large DNA clone can be digested into a number of smaller fragments which can be separated on an agarose gel and transferred to nitrocellulose paper by the Southern blotting technique. Fragments of the paper containing each separated band can then be cut out and used for hybrid selection. In this way, it is possible to identify regions within a large clone which encode a specific mRNA and to distinguish these regions from the coding regions of contiguous genes (as in the case of viral genomes). In the case of genomic clones, this technique can be used to distinguish exons from introns. Many of these applications are dealt with in detail in a recent review.[15]

The final section of this chapter details an example of how the HST procedure has been adapted to the study of a specific genetic problem in the field of hematology: the characterization of human hemoglobin mutants.

Characterization of globin structural mutants using the technique of hybrid-selected translation[17]

Over 150 α-globin and 120 β-globin structural mutants are now documented in the literature. These mutants are defined at the level of protein structure and many have characteristic patterns of migration on a variety of analytic gels and columns. However, when faced with a mutant which has not been previously identified, or a mutant which cannot be specifically defined by gel analysis, definition depends upon isolation of the abnormal globin protein and analysis of its peptide fragments. Such analysis is routinely carried out by reference laboratories. An additional complication arises in the characterization of α-globin mutants. As there are two α-globin genes in the human genome, which both encode an identical α-globin, the protein analysis approach cannot define which of the two α-globin genes encodes any particular α-globin mutant. The technique of hybrid selection has been applied to this problem to fill this gap. As shown below and diagrammed in Figure 12.2, this approach provides a rapid and definitive localization of globin structural mutants to either the α- or β-globin gene and, in the case of α-globin mutants, it can specifically localize a mutant to either the α1 or α2 locus. In order to use the HST approach for this problem one needs DNA clones which will specifically hybridize to the three mRNAs in question, β, α1 and α2. β-globin mRNA can be specifically hybrid-selected using any reasonably sized β-globin cDNA clone (for example the near full-length β-globin cDNA, pSAR6 — Fig. 12.2). Separation of the α1 and α2-globin mRNAs is more difficult as these two mRNAs are identical in structure from their 5' terminus through to the termination codon and only diverge structurally in the 3' non-translated region. A full-length α-globin cDNA (pMC18 — Fig. 12.2) will hybridize to both of the α-globin mRNAs. Therefore, to isolate these two mRNAs separately, their divergent 3' non-translated regions had to be subcloned. By using these two α1 and α2 3' non-translated regions subclones

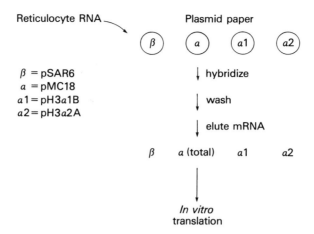

Fig. 12.2 Scheme for locus identification of globin structural mutants by hybrid selected translation. Total reticulocyte RNA is hybridized to nitrocellulose paper containing plasmids which will specifically hybridize to β-, α-, α1-, or α2-globin mRNA. After hybridization and extensive washing, the specifically bound mRNA is eluted from the paper and translated *in vitro* in the presence of [^{35}S]methionine. Analysis of the labeled protein products encoded by each of the mRNA samples reveals the identity of the mutant globin protein.

(pH3α1B and pH3α2A in Fig. 12.2) which diverge from each other by 16% in primary structure, and by adjusting the hybridization conditions so that they will specifically hybridize to their homologous mRNAs, the two α-globin mRNA species can be individually isolated. Following the scheme in Figure 12.2, we can isolate from a single sample of mRNA all three globin mRNAs, β, α1, and α2. By translating each of these mRNAs *in vitro*, and analyzing the synthesized products, the identity of the mutant globin chain can be determined. An example of such a study is the locus assignment of the α-globin mutant Hb Hasharon ($α^{47His}$) shown in Figure 12.3 and detailed in reference [17]. The mutant in this case is assigned to the α2 globin gene since the mutant protein is produced only by the α2 mRNA. The only alternative method presently available to assign α-globin structural mutants to one of the two α-globin genes is to clone and sequence the genes encoding each α-globin mutant. This use of *in vitro* translation to specifically assign α-globin mutants to the α1 or α2 gene locus, therefore demonstrates how *in vitro* translation techniques can substantially facilitate the analysis of a genetic problem.

REFERENCES

1 Pelham H R B, Jackson R B. An efficient mRNA-dependent translation system from reticulocytes lysates. Eur J Biochem 1976; 67: 247–257.
2 Merrick W C. Translation of exogenous mRNA in reticulocyte lysate. Methods Enzymol 1983; 101: 606–615.
3 Anderson C W, Straus J W, Dudock B S. Preparation of a cell-free protein-synthesizing system from wheat germ. Methods Enzymol 1983; 101: 635–644.

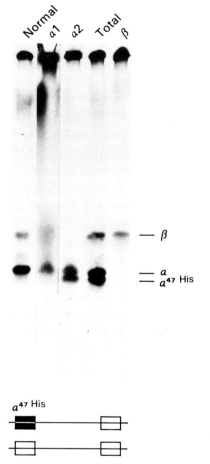

Fig. 12.3 Localization of the α-globin mutant $\alpha^{47\ His}$ to the α2 gene locus. Hybrid-selected α1-, α2-, and β-globin mRNA from the reticulocyte RNA of an individual with the Hb Hasharon mutation ($\alpha^{47\ His}$) was translated *in vitro* and analyzed by TAU electrophoresis. The normal four α-globin gene organization diagrammed at the bottom was determined by Southern blot analysis. The localization of the Hasharon mutation to the α2 locus (the more 5′ or leftward of the two adjacent α-globin loci) was determined by the presence of the mutant protein in the analysis of labeled *in vitro* translation products of the α2-globin mRNA.

4 Brown G D, Peluso R W, Moyer S A, Moyer R W. A simple method for the preparation of extracts from animal cells which catalyze efficient in vitro protein synthesis. J Biol Chem 1983; 258: 14309–14314.
5 Gurdon J B, Lane C D, Woodland H R, Marbaix G. Use of frog eggs and oocytes for the study of messenger RNA and its translation in living cells. Nature 1971; 233: 177–182.
6 Marbaix G, Huez G. In: Celis J E, Graessmann A, Loyter A, eds. Transfer of cell constituents into eucaryotic cells. New York: Plenum, 1980.
7 Moldave K Eukaryotic protein synthesis. Annu Rev Biochem 1985; 54: 1071–1109.
8 Lingrel J B, Borsook H. A comparison of amino acid incorporation into the hemoglobin and ribosomes of marrow erythroid cells and circulating reticulocytes of severely anemic rabbits. Biochemistry 1963; 2: 309.

9 Scheele G, Dobberstein B, Blobel G. Transfer of protein across membranes. Eur J Biochem 1978; 82: 593–599.
10 Walter P, Blobel G. Purification of a membrane-associated protein complex required for protein translocation across the endoplasmic reticulum. Proc Natl Acad Sci USA 1980; 77: 7112–7116.
11 Liebhaber S A, Kan Y W. Different rates of mRNA translation balance the expression of the two human α-globin loci. J Biol Chem 1982; 257: 11852–11855.
12 Maniatis T, Fritsch E F, Sambrook J. Molecular cloning, a laboratory manual. 1982; p 191–197.
13 Wreschner D, Herzberg M. A new blotting medium for the simple isolation and identification of highly resolved mRNA. Nucleic Acids Res 1984; 12: 1349–1359.
14 Hastie N D, Held W A. Analysis of mRNA populations by cDNA-mRNA hybrid-mediated inhibition of cell-free protein synthesis. Proc Natl Acad Sci USA 1978; 75: 1217–1221.
15 Miller J S, Patterson B M, Ricciardi R P, Cohen L, Roberts B E. Methods utilizing cell-free protein-synthesizing systems for the identification of recombinant DNA molecules. Methods Enzymol 1983; 110: 650–674.
16 Liebhaber S A, Cash F E, Shakin S H. Translation associated helix-destabilizing activity in rabbit reticulocyte lysate. J Biol Chem 1984; 259: 15597–15602.
17 Liebhaber S A, Cash F E. Locus assignment of α-globin structural mutations by hybrid-selected translation. J Clin Invest 1985; 75: 64–70.

13
Transfection of DNA into mammalian cells
M. Donovan-Peluso   A. Bank

INTRODUCTION

Techniques for the introduction and stable integration of foreign DNA sequences into mammalian chromatin have developed rapidly during the last several years. These strategies have enabled investigators to ask specific questions about the regulation of these genes in cells. These experiments have defined, to some extent, both the role of chromosomal position on gene expression and the amount of flanking sequences upstream and downstream (called cis-acting elements) of coding regions required for expression of individual genes.[1-5] Using this approach, questions have begun to be asked regarding the interactions of cis-acting elements involved in modulating gene expression[6-16] with trans-acting factors present in cells.[17-20]

In addition to their biological interest these techniques have also demonstrated the feasibility of transferring genes to recipient cells with high frequency and high level expression.[21-34] The possibility of using these methods for gene therapy or autotransplantation in patients with genetic diseases by inserting functional genes into cells is coming closer to reality.[33,35,36]

In this chapter the strategies available for the identification and selective growth of cells containing transfected genes will be discussed. Genes that provide a selective advantage to cells that integrate and express them are called 'selectable' genes. We will describe the use of selectable genes such as thymidine kinase (TK)[22,24,37-39] and adenine phosphoribosyl transferase (APRT),[40] two genes that require cells deficient in these activities. In addition, we will discuss 'dominant selectable' genes that can be used in all cell types including bacterial xanthine-guanine phosphoribosyl transferase (XGPRT),[41-43] bacterial aminoglycoside 3′ phosphotransferase (Neo^r),[44,45] bacterial hygromycin B phosphotransferase (HPH),[46] and murine dihydrofolate reductase (DHFR).[47-51]

The techniques available for introducing DNA into mammalian cells will be assessed. These include calcium phosphate precipitation,[21-24] protoplast fusion,[14,25,26] electroporation,[27-29] and the use of retroviruses.[30-34] Finally, protocols for the analysis of transfected cells including growth and harvesting of cells,[52] analysis of integrated DNA,[53,54] and analysis of RNA transcripts produced by transfected genes[55-61] will be discussed.

SELECTION STRATEGIES

Complementation of cells deficient in an activity

TK selection[22,24,37–39]

The ability to utilize TK selection requires that the host cells do not possess their own functional copy of the TK gene (TK−) and cannot convert thymidine to inosine monophosphate (IMP). Selective pressure to maintain the TK− phenotype is exerted by adding bromodeoxyuridine to the growth medium. Cells that are TK− cannot metabolize this potentially lethal analogue into a DNA precursor and, therefore, the cells are viable. Transfected TK+ cells are selected in medium containing hypoxanthine (10^{-4} M), aminopterin (2×10^{-5} M), and thymidine (10^{-4} M) (HAT medium). Aminopterin blocks *de novo* nucleotide biosynthesis, hypoxanthine is the source of purines, and thymidine provides the substrate for pyrimidine biosynthesis.

APRT selection[40]

Medium containing adenine analogues such as 2,6 diaminopurine or 8-azaadenine is used as selection against APRT+ cells. The APRT− variant of TK− cells grows in the presence of diaminopurine. Selection for APRT+ cells is growth in medium containing azaserine (0.05 mM) to block *de novo* purine biosynthesis and adenine (0.1 mM) which can be used for purine nucleotide synthesis.

Dominant selection

More recently, genes that can be introduced into any cell type and confer resistance to a variety of agents have become available. These genes have the advantage that special cells deficient in an activity are not required.

XGPRT selection[41–43]

This selection interferes with the *de novo* guanosine synthesis pathway and takes advantage of the difference in sensitivity of the mammalian hypoxanthine phosphoribosyl transferase (HPRT) and the bacterial XGPRT enzymes to inhibition by mycophenolic acid (MPA). The mammalian enzyme is only weakly able to convert xanthine to xanthine monophosphate (XMP). In guanine-free media, addition of aminopterin (2 µg/ml) to block the conversion of precursors to IMP, and MPA (25 µg/ml) to block the conversion of IMP to XMP effectively blocks guanosine synthesis by mammalian HPRT. Transfected cells containing the bacterial XGPRT gene can be selectively grown in this medium when xanthine is supplied as the sole precursor for guanosine synthesis.

Neo^r selection[44,45]

Although eukaryotic cells are insensitive to neomycin and kanamycin, an analogue of these compounds, G418 or Geneticin, has been developed that inhibits protein synthesis and is toxic for eukaryotic cells. The Neo[r] gene encodes a phosphotransferase that inactivates the G418 and, therefore, cells containing this gene preferentially survive. Since large variations in cellular susceptibility to the toxicity of G418, as well as variations in the potency of different batches of G418 have been observed, it is imperative that the dose be adjusted for each particular line. Reported selective concentrations range between 0.2 mg/ml and 1 mg/ml. We have found the Neo[r] gene and G418 selection extremely useful because of the lack of spontaneous G418-resistant cells. Thus, when G418-resistant cells are identified after transfer of Neo[r] genes, these cells invariably contain the Neo[r] gene and other co-transferred genes.

HPH selection[46]
Hygromycin B is an antibiotic that is lethal for bacterial, lower eukaryotic, and mammalian cells. It has been shown to interfere with protein synthesis by interfering with ribosomal translocation and aminoacyl-tRNA recognition. Transfection of eukaryotic cells with the HPH gene confers resistance to hygromycin B in concentrations ranging from 0.05 to 0.4 mg/ml of medium.

DHFR selection[47-51]
Tetrahydrofolate is involved in the synthesis of DNA, RNA, and protein. DHFR (dihydrofolate reductase) is the enzyme that catalyzes the conversion of inactive dihydrofolate to tetrahydrofolate, the biologically active form. Folate antagonists such as methotrexate (MTX) have a higher affinity for the binding site of DHFR and prevent its biological activity, resulting in cell death. A cloned DHFR cDNA with a reduced affinity for MTX ($\frac{1}{270}$ of wild type) is available for use as a dominant selectable marker. When introduced into cells it confers resistance to 50–500 nM MTX in selective medium lacking hypoxanthine, glycine, and thymidine (HGT). Concentrations of MTX as high as 0.1 mM yield colonies, although fewer in number. A second cloned DHFR gene encoding the wild type enzyme has been used to successfully amplify both itself as well as co-transfected genes by progressively increasing the dose of MTX (0.01, 0.02, 0.1, 1.0, and 50 μM) in HGT.

METHODS OF GENE TRANSFER

Calcium phosphate precipitation[21-24]
One day prior to transformation, cells are plated at a subconfluent density of 0.5×10^6 cells/10 cm dish in medium lacking $CaCl_2$ such as Dulbecco's Modified Eagles Medium (DME). Medium containing $CaCl_2$, such as RPMI, forms huge, virtually insoluble precipitates when the DNA is subsequently added, and should be avoided. DNAs are routinely precipitated from aqueous stock solutions by ethanol precipitation (see Chs. 3,4,5) in amounts required for 10 plates per experimental group. Usually five plates per group are used as carrier controls. Typical amounts of nucleic acid that we have used in co-transformation exper-

iments include 10 μg/plate of carrier salmon sperm DNA, 10 μg/plate of non-selected plasmid containing the gene to be analyzed and 5 μg/plate of the plasmid containing the selectable gene. For convenience the genes whose expression is to be analyzed can be subcloned (Chs. 5–7) directly into the selectable plasmid and used to transform eukaryotic cells. In this case 10 μg/plate of carrier DNA and 15–25 μg/plate of plasmid DNA are used.

Salmon sperm DNA is precipitated under sterile conditions in 100 μg aliquots, resuspended in 0.5 ml of 10 mM Tris, pH 7.5, 0.1 mM EDTA (TE) and resuspended for 2–3 days at 4°C. Plasmid DNA is precipitated under sterile conditions and resuspended in 0.5 ml of TE. For 10 plates/group 10 ml of precipitate will be required.

Two tubes/group are prepared: one contains 5 ml of 2 × HEPES buffered saline (HEBS) brought to pH 7.12 and sterile filtered immediately prior to use. (2 × HEBS = 1.6% NaCl, 0.074% KCl, 0.025% $Na_2HPO_4.2H_2O$, 0.2% dextrose, 1.0% HEPES (N-2-hydroxyethyl piperazine-N′-ethane sulfonic acid (Sigma)), pH 7.12. 50 ml aliquots of 2 × HEBS may be stored frozen (−20°C). After thawing, the pH should be readjusted and the solution sterile filtered prior to use. A second tube containing 2.5 ml of 0.5 M $CaCl_2$ and the sterile, resuspended DNA in a final volume of 2.5 ml is mixed and the contents of the DNA.$CaCl_2$ tube are slowly added (dropwise) to the HEBS tube with continuous bubbling and allowed to sit at room temperature 20–30 minutes to allow the DNA–$CaPO_4$ precipitate to form. The precipitate should be very finely granular.

1 ml of precipitate suspension is added to each plate of subconfluent cells containing 10 ml of growth medium and swirled to mix. Following the addition of the suspension the medium should have an opalescent, cloudy appearance. After the cells have remained in contact with the precipitate for 18–24 hours, replating of the cells onto fresh plates containing fresh medium is critical, since sustained contact with the $CaPO_4$ is toxic for the cells. After a further 24 hours, the selective compound is applied and the medium is changed every 4 days. Once the cells on the control plates have died and small colonies are visualized on the experimental plates, the living and dead cells may be separated on Lymphocyte Separation Medium (LSM; Bionetics Laboratory Products, Kensington MD).

LSM is a Ficoll solution that separates cells according to density. After centrifugation, live cells are found at the cell suspension:LSM interface, dead cells are present in the pellet. In a 15 ml conical centrifuge tube 3–5 ml of LSM is layered with 10 ml of medium (from one plate) containing cells and centrifuged at 1500 rpm for 15–20 minutes in a swinging bucket rotor at room temperature. The medium is aspirated to within 1–2 cm of the interface. Using a sterile pipet approximately 2–3 ml containing the living cells is removed to a new centrifuge tube, washed with 10 ml of PBS and resuspended in fresh medium. Colonies should be visible within 4–5 days.

This method of gene transfer requires large amounts of plasmid DNA, is labor intensive and is relatively inefficient. For certain cell lines the efficiency can be so low as to be prohibitive. It is reproducible, however, and many investigators have used it successfully to introduce genes into eukaryotic cells.

Protoplast fusion[25,26]

The protoplast fusion method is favored by many laboratories for gene transfer because it is technically simple and potentially offers higher efficiencies of gene uptake by the target cells. Bacterial cells containing the plasmids one wishes to transfect are treated with lysozyme to remove cell walls, creating protoplasts. The protoplasts are then incubated with the host in the presence of an agent (polyethylene glycol) that causes the protoplasts and the target cells to fuse, thus allowing the plasmid DNA to enter the target cell.

Growth of bacteria and preparation of protoplasts

All solutions used in the preparation of protoplasts should be sterile. 15 ml bacterial overnights are grown in 100 ml capped bottles at 37°C in N-Medium (0.1% glucose, 0.1% yeast extract, 1.0% bactotryptone, 0.8% NaCl, 0.03% $CaCl_2$) containing antibiotic, using sterile conditions and inoculated under a hood. The next day the overnight culture is diluted and *E. coli* containing plasmids are grown to OD A_{600} 0.6–0.8 in sterile, capped Gibco bottles containing 100 ml of media plus the appropriate antibiotic. When the desired density is reached, 10 ml of 1.5 mg/ml sterile chloramphenicol is added (final concentration is 0.15 mg/ml) and the culture is shaken overnight (12–16 hours) to allow the plasmids to amplify.

The bacteria are harvested by centrifugation at 7000 rpm for 5 minutes in sterile centrifuge bottles. The supernatant is decanted and the pellet is resuspended in 5 ml of chilled 20% sucrose, 50 mM Tris, pH 8, and 1 ml of fresh *sterile* lysozyme (5 mg/ml in 250 mM Tris, pH 8) and incubated on ice 5 minutes. 2 ml of 250 mM EDTA is added, mixed by gentle swirling and incubated on ice 5 minutes. 2 ml of 50 mM Tris, pH 8, is added *dropwise*, swirled to mix, and placed at 37°C for 8.5 minutes. This suspension is slowly diluted with 40 ml DME without serum containing 10% sucrose, 10 mM $MgCl_2$, mixed gently, and incubated at room temperature for 10 minutes. We have used protoplasts for up to 90 minutes after preparation.

Fusion with eukaryotic cells

Protoplast and eukaryotic cell fusion is mediated by polyethylene glycol 1500 MW (PEG) purchased from Gallard Schlesinger Chemical Manufacturing, 584 Mineola Ave, Carle Place, N.Y. 2×10^6 to 1×10^7 cells/tube are spun down in a 50 ml centrifuge tube, washed with $1 \times$ DME without serum, centrifuged, and the supernatant aspirated, leaving the pellet. 5 ml of protoplast suspension is added to the resuspended pellet and the mixture is centrifuged for 5 minutes at 1600 rpm. The supernatant is aspirated, the pellet is *resuspended*, and 0.5 ml PEG:DMSO (prepared by combining 49 ml DME without serum, 50 g PEG, 6 ml 2 M Tris, pH 8.0, sterile filtering and adding 15 ml DMSO (dimethyl sulfoxide) is added and the suspension shaken gently for one minute followed by the addition of 0.5 ml 25% PEG (prepared by combining 25 g PEG, 65 ml

DME without serum, and 5 ml of M Tris, pH 8, and sterile filtering) and shaking gently for 2 minutes. This suspension is immediately diluted with the addition of 9 ml of DME without serum, mixed, centrifuged, and the supernatant is aspirated. The pellet is resuspended in 12 ml of growth medium, pipeted well to mix, and plated immediately on 96 well plate at 10^4 cells per well for selection of stable transformants, or grown in flasks for transient expression assays. Cells are allowed to recover for one day before application of the selective agent. We use half of the maximal dose of selective agent for several days and then increase to the maximal dose, which is determined empirically.

We have used protoplast fusion to dramatically increase the efficiency of gene transfer into erythroleukemia (K562) cells. However, not all cell types exhibit the same response. This technique, although not difficult, requires practice, and care must be taken at each step to prevent bacterial contamination. Finally, the success of these gene transfer experiments requires that the bacteria are converted to spheroplasts and not lysed, and problems at this step are not infrequent.

Electroporation[27-29]

Electroporation employs the principle that a short, high voltage (1000–4000 V/cm^2) pulse of electricity applied to eukaryotic cells will perturb the cellular membrane sufficiently to enable nucleic acid present in the buffer to diffuse into the cells. It is important to check viability and determine the appropriate voltage that gives 50% viability for each cell type. One critical requirement for this technique is a power supply with large capacitors (i.e. an ISCO Model 494 dc power supply) or one that can be modified to provide such capacitance. Prototype electroporation devices are now available commercially.

Actively growing cells at a density of 0.5×10^6/ml of medium are centrifuged, suspended in ice-cold phosphate buffered saline (PBS = 0.02% KCl, 0.02% KH_2PO_4, 0.8% NaCl, 0.114% Na_2HPO_4) without added Mg^{2+} or Ca^{2+}, centrifuged again and the pellet resuspended in 0.65 ml of ice-cold PBS and transferred to sterile 'shocking chambers' and wrapped with sterile parafilm. One type of 'shocking chamber' available from BioRad consists of sterile flat-sided plastic cuvettes with aluminum foil electrodes epoxied to the 2 non-transparent sides. Linearized or supercoiled DNA at concentrations of 10–20 μg/ml is added to the cells in PBS, and this mixture is incubated in the cuvette on ice for approximately 5 minutes. The cuvettes containing cells and DNA are mixed and transferred to the shocking device that is precooled to 4°C, the pulse is delivered, and the cells and DNA are allowed to sit on ice for an additional 5 minutes. Foaming at the top of the cuvette indicates some cell lysis and is acceptable. The cells are removed to a 50 ml centrifuge tube (suspension will be viscous), the cuvette is rinsed with additional medium, and the cells are resuspended in a final volume of 50 ml and plated (0.5 ml/well, 0.5×10^5 cells/well) on 96 wells (4 plates) of 24 well limbro dishes. The percent viability following transfection is calculated using trypan blue exclusion, and this number used to adjust the plating as well as to provide estimates of the success and efficiency of individual transfer experiments. In our experience, 50–80% viability is most compatible with successful gene transfer.

The major advantage of this technique is that it facilitates the high efficiency transfer of nucleic acid into a broad range of cell types. This is particularly important when working with cell lines that exhibit an extremely low efficiency with alternative methods. Another advantage is the speed and convenience of this method. It literally takes minutes to transfect the cells and this is a major advantage, especially when several constructs are to be transfected.

Retroviruses[30-34]
Retrovirus transduction of genes into eukaryotic cells offers the advantages of high efficiency of infection, single or low copy number integration, and a defined integrated provirus structure. The availability of efficient gene transfer methodologies is critical if the prospects of infecting a low abundance precursor cell for bone marrow mediated gene therapy are to become a reality. Using retroviruses, genes have been transduced into progenitor cells and these cells have been used to successfully engraft lethally irradiated recipient mice.

Although there are size constraints that must be considered when constructing recombinant retroviruses, the use of small transcription units and cDNAs has allowed several genes to be successfully transferred and expressed in hemopoietic progenitor cells. The goal for gene therapy is to achieve gene transfer into primitive stem cells and reintroduce these cells into appropriate hosts.

Retroviral gene transfer methods are elaborate in the sense that facilities for maintenance of viral stocks and host cells must be available. The reader is referred to references [30-34] for descriptions of these techniques.

ANALYSIS OF CELL LINES CONTAINING TRANSFECTED GENES

Harvesting of transfected cells[52]
Once transfected cells are identified by selection, they are grown in mass culture. Usually 50×10^6 cells per flask are harvested, and DNA and RNA are prepared for subsequent analysis. Cells are centrifuged for 5 minutes at 1500 rpm, the supernatant decanted, and the pellets washed with 20 ml of PBS followed by centrifugation and decanting of the supernatant. The cell pellets are resuspended in 5 ml of 20 mM Tris, 10 mM EDTA and allowed to swell on ice for 3–5 minutes. 0.25 ml of 20% sarkosyl is added to lyse the cells, followed by 5.25 g of CsCl which is gently dissolved, layered on a 3 ml 1.35 g/ml CsCl cushion in an SW41 tube, and centrifuged for 16 hours at 30 000 rpm, at 20°C (see Ch. 3).

After centrifugation, the top of each gradient is removed and discarded until a very viscous layer, near the interface, containing DNA is reached. This layer is removed and saved for DNA analysis. The remaining liquid in the tube is decanted, and the tubes are allowed to drain. A visible pellet in the bottom of each tube contains CsCl and RNA. 1 ml of RNA resuspension buffer (10 mM Tris, pH 7.5, 1 mM EDTA, 0.2% SDS) is added to each tube and the pellets are allowed to dissolve. Each pellet is dissolved in a final volume of 4.5 ml of buffer, 0.5 ml of 3 M sodium acetate, and 2.5 volumes of ethanol are added to each tube and RNA is precipitated overnight at $-20°C$. The RNA is centrifuged, the pellet is washed with 80% ethanol, dried and resuspended in 0.3 ml of 10 mM Tris, pH 7.5, 1 mM EDTA, 0.1% SDS. The absorbance A_{260} is read to determine the RNA yield.

To remove cesium chloride from the DNA it is dialyzed against 2 changes of 500 volumes each of 10 mM Tris, pH 7.6, 1 mM EDTA, 0.1% SDS at room temperature, followed by phenol extraction and ethanol precipitation.

Analysis of DNA

Southern blotting and hybridization[53-54]

To analyze the specific gene content of cellular DNA, Southern blot analysis is used (see also Ch. 2). The DNA is digested to completion with restriction enzymes run on a 0.75% agarose gel containing 0.75 μg/ml ethidium bromide at 20 V for 16 hours. Following electrophoresis the gel is photographed and the DNA is denatured *in situ* in 500 ml of 1 M potassium hydroxide for 25 minutes on a platform shaker and neutralized with two changes of 1 M Tris, pH 7.0, each approximately 500 ml. The gel is equilibrated with 200 ml of 6 × SSC (20 × SSC = 3 M NaCl; 0.3 M Na Citrate pH 6.8 with M citric acid) for 20 minutes. A nitrocellulose filter is cut to the dimensions of the gel, washed with distilled water, and equilibrated for 30 minutes in 6 × SSC. 800 ml of 6 × SSC is placed in a dish containing a plexiglass platform and a Whatman 3 mm wick. The gel, nitrocellulose filter, and 2 pieces of Whatman 3 mm paper are placed on top of the wick. A stack of paper towels is placed on top and the DNA is allowed to transfer for 16 hours.

The filter is baked for 3–5 hours at 80°C in a vacuum oven and stored in a vacuum desiccator until used. The filter is pre-hybridized in 300 ml of 1 × phosphate (3 × = 900 ml of 0.75 M Na_2HPO_4, 0.1 M Na pyrophosphate plus 600 ml of 0.75 M NaH_2PO_4; pH 7.0), 6 × SSC, 0.02% BFP (B = bovine serum albumin, F = Ficoll, P = polyvinylpyrrolidone), 50 μg/ml salmon sperm DNA, for 4 hours at 68°C. The amount of hybridization solution is calculated as 0.02 ml/cm² of filter. The probe (2 × 10⁶ cpm/ml) (See Chs. 4 and 5 for probe preparation) is added to the appropriate amount of 0.02% BFP, 0.001 M EDTA, 0.1% SDS, 50 μg/ml salmon sperm DNA and is boiled for 5 minutes and chilled on ice for 5 minutes. SSC is added to a final concentration of 2 × and this solution is placed in the hybridization bag which is carefully sealed to remove any air bubbles and incubated for 24 hours at 68°C.

The filter is washed the next day in 6 one-liter washes of 2 × SSC, 0.5% SDS for 10 minutes each at 68°C, 1 one-liter wash in 0.1 × SSC, 0.5% SDS for 10 minutes at 68°C, and 1 one-liter wash in 2 × SSC for 5 minutes at room temperature. The filter is air dried and autoradiographed with Kodak XR2 film at $-70°C$ with a Dupont Lightning-Plus XI intensifying screen for 24–48 hours.

Analysis of RNA

RNA-dot blots
RNA is spotted directly onto Biodyne A membranes* (Pall Ultrafine Corp., Glen Cove, New York) in multiple aliquots of 1.5 μl, allowing the spot to dry between applications. The blot is air dried and baked at 80°C in a vacuum oven for 1–2 hours, and pre-hybridized in a plastic bag at 42°C for 1–2 hours in 4 ml of 5 × Denhardt's (100 × = 2% Ficoll, 2% polyvinyl Pyrrolidone, 2% bovine serum albumin), 5 × SSC, 50 mM Na phosphate, pH 6.5, 0.1% SDS, 250 μg/ml salmon sperm DNA and 50% formamide. Hybridization solution is identical to the pre-hybridization solution except that 2×10^6 cpm/ml of nick translated probe is added, the solution is heated for 5 minutes at 95°C to denature the probe and cooled for 5 minutes on ice prior to addition to the filter. A hybridization volume of 2 ml per 100 cm^2 of membrane is used. The filters are hybridized at 42°C overnight.

Following hybridization the membranes are dipped briefly into wash buffer (2 × SSC, 0.1% SDS), sealed in a fresh bag containing 250 ml of wash buffer per 100 cm^2, agitated for 5 minutes at room temperature on a platform shaker, and this step is then repeated. The membrane is removed, placed in a fresh bag containing 250 ml per 100 cm^2 of 0.1 × SSC, 0.1% SDS heated to 50°C, incubated at 50°C for 15 minutes, and this wash is repeated. The membrane is blotted briefly, wrapped in Saran wrap and autoradiographed. If the probe is to be subsequently removed it is important not to allow the membrane to dry out completely.

Hybridized probe is removed by incubating the membrane in 100 ml of 10 mM Na phosphate, pH 6.5, 50% formamide per 100 cm^2 at 65°C for 1 hour. The membrane is washed once in 250 ml of 2 × SSC, 0.1% SDS per 100 cm^2 for 15 minutes at room temperature, agitating on a platform shaker. To ensure that all the probe is removed the membrane is blotted briefly, wrapped in Saran wrap and autoradiographed. To rehybridize the blot it is pre-hybridized and hybridized as described above.

Northern blotting and hybridization (RNA)[55–56] (see also Chs. 3 and 11)
10 × morpholinopropane sulfonic acid (Mops) buffer (200 mM Mops (Sigma), 50 mM sodium acetate, 10 mM EDTA, pH 7.0) is prepared fresh, filter sterilized, stored at −20°C and used within one week. To prepare the gel, 1 g of agarose is dissolved in 73 ml of dH$_2$O with boiling, 10 ml of 10 × Mops buffer and ethidium bromide to 0.75 μg/ml is added, and the gel is cooled to 60°C. Formaldehyde is added to 2.2 M final (16.9 ml of 30% formaldehyde (Fisher)) and the gel is poured and solidified under a hood. Running buffer is 1 × Mops.

RNA is lyophilized and resuspended in 16.4 μl of 50% formamide, 1 × Mops, and 3.6 μl of formaldehyde is added to make 2.2 M final. The samples are denatured at 65°C for 10 minutes and chilled on ice for 5 minutes. 5 μl of 20%

---

* Editor's note: nitrocellulose sheets will also suffice. Several companies sell devices that hold the filter in position under wells or slots onto which the RNA can be spotted.

Ficoll, 50 mM EDTA, 0.1% bromophenol blue is added to each sample and they are loaded onto the agarose gel and electrophoresed at 20 V overnight. The gel is photographed and positions of the ribosomal bands are recorded as size markers.

The gel is equilibrated to 10 × SSC by soaking in 100 ml of 20 × SSC (3 M NaCl, 0.3 M sodium acetate, pH 6.8, with M citric acid) for 60 minutes. Excess gel is trimmed and the size is measured. One piece of nitrocellulose (0.45 μm, Scheleicher and Schuell), two pieces of Whatman 3 mm, and a 20 cm stack of paper towels are cut to the size of the gel, two pieces of Whatman 3 mm are cut to 18 cm × 33 cm for the wick. The nitrocellulose filter is soaked in freshly boiled dH_2O for 5 minutes and then 10 × SSC for 15 minutes.

500 ml of 10 × SSC is placed in a dish containing a plexiglass platform, and the wet Whatman wick is layered over the platform with both ends immersed in the buffer. The gel, nitrocellulose filter, and 2 pieces of Whatman 3 mm paper are placed on top of the wick. A stack of paper towels is placed on top, and the RNA is allowed to transfer for 16 hours. The nitrocellulose filter is removed, rinsed, briefly in 3 × SSC, air dried, and baked at 80°C under vacuum for 5 hours. Filters are stored at room temperature in a vacuum desiccator.

The filter is placed in a sealable bag and sealed as close to the filter as possible to minimize air space. 10 ml of pre-hybridization solution (50% formamide, 10% dextran sulfate, 5 × SSC, 4 × Denhardts (100 × Denhardts = 2% Ficoll, 2% polyvinylpyrrolidone, 2% bovine serum albumin), 0.1% SDS) is added to the bag, which is carefully sealed, taking care to remove all air bubbles. The filter is pre-incubated at 42°C for 5 hours. The hybridization solution (20 $\mu l/cm^2$) containing 50% formamide, 10% dextran sulfate, 0.1% SDS, 1 mM EDTA, 0.1 mg/ml salmon sperm DNA, and 2×10^6 cpm/ml of probe is boiled for 5 minutes and quenched on ice for 5 minutes. SSC is added to a final concentration of 2 ×. The pre-hybridization solution is removed from the bag, eliminating as much liquid as possible, and replaced with hybridization solution. Care is again taken to reseal the bag, removing all air bubbles, and the filter is hybridized at 42°C for 16–48 hours.

Following hybridization, the filters are washed in solutions of increasing stringency to remove any non-specific hybridization. The first wash is 500 ml of 2 × SSC at room temperature for 20 minutes. All other washes are at 68°C for 10 minutes and consist of 2 × SSC, 1 × SSC, 0.5 × SSC, 0.2 × SSC, 0.1 × SSC. Following the washes the filter is air dried and autoradiographed at −90°C with a Kronex Lightning-Plus intensifying screen.

To re-hybridize a Northern blot with a different probe, the old probe is washed off in 500 ml of 5 mM Tris, pH 8, 0.2 mM EDTA, 0.05% sodium pyrophosphate, and 0.1 × Denhardt's for 2 hours and 68°C. To ensure that the unwanted probe is no longer present, the blots are autoradiographed and the lack of signal is documented.

Sp6 RNase protection[57]

Sp6 probes are synthesized by a modification of Melton et al (see also Ch. 4, 10). 1–3 μg of linearized template is incubated in a reaction buffer containing 40 mM

Tris, pH 7.5, 10 mM NaCl, 6 mM $MgCl_2$, 2 mM spermidine, 0.5 mM each of ribonucleotides ATP, CTP, GTP, and 0.2 mM cold UTP, 50–100 μCi of labeled UTP (Amersham), fresh dithiothreitol to 16 mM, 36.5 units of placental ribonuclease inhibitor (Rnasin, Boehringer Mannheim (BM)), and 30 units of Sp6 polymerase (BM), and incubated for 90 minutes at 40°C. DNase (BRL, RNase free) is added to 40 μg/ml with an additional 36 units of Rnasin and incubated at 37°C for 15 minutes, followed by phenol extraction and chromatography on a G100 Sephadex column prepared in a 5 inch Pasteur pipet equilibrated with 10 mM Tris, pH 7, 1 mM EDTA, and 0.1% SDS. Ten fractions are collected, fraction 1 contains 300 μl, fraction 2 contains 200 μl, and fractions 3–10 contain 100 μl each. The peak fractions are pooled, yeast tRNA (10 μg) is added as carrier, sodium acetate to 0.3 M and 2.5 volumes of ethanol, and precipitated. The pellet is resuspended in 80% formamide, 40 mM Pipes, pH 6.4, 400 mM NaCl, 1 mM EDTA (80% FAHB). After calculation of recovery, the probe is diluted to give a final concentration of 5 ng/20 μl in FAHB, and 20 μl/sample is hybridized to 10 μg of total RNA at 50°C for 20 hours.

The probe:mRNA hybrids are diluted with 300 μl of 10 mM Tris, pH 7.5, 5 mM EDTA, 300 mM NaCl containing RNase A (Sigma type X-A) to 40 μg/ml and RNase T1 (Sigma, Rl003) to 2 μg/ml, incubated at 37°C for 60 minutes, followed by addition of SDS to 0.6% and proteinase K to 160 μg/ml and incubated at 37°C for 15 minutes. After phenol extraction and precipitation, the pellets are resuspended in denaturing loading dyes (90% formamide, 1 mM EDTA, 0.05% xylene cyanol, 0.05% bromophenol blue), are denatured at 95°C for 5 minutes, quick chilled and electrophoresed at 2000 V:24 mA on 6–8% acrylamide:7 M urea sequencing type gels.

S_1 analysis[58-60]

Uniformly labeled probes synthesized on M13 single strand (ss) templates and strand separated, or end labeled probes (see Chs. 4, 5, 11) may be used for S_1 analysis. Precipitated probes are resuspended in enough 80% FAHB to guarantee 10 μl/RNA hybridization with 0.01 to 0.1 pM of probe per reaction. Whenever a new probe is used, control hybridizations are done to determine empirically whether or not a given number of picomoles of probe will provide probe excess for a known amount of control RNA. The RNA (20 μg) is lyophilized and resuspended in 10 μl of probe, vortexed to mix, denatured at 68°C for 10 minutes and hybridized at a temperature approximately 5°C above the melting temperature to favor the generation of RNA:DNA hybrids for 20–40 hours.

Following hybridization, 110 μl of S_1 mix (200 mM NaCl, 50 mM sodium acetate, pH 4.5, 1 mM zinc sulfate, 20 μg/ml denatured salmon sperm DNA and 10 000–20 000 units/ml of S_1 nuclease) prewarmed to 37°C, is added to each tube, vortexed, and incubated at 37°C for 60 minutes. Phenol (100 μl/tube) is added, vortexed, centrifuged, and the aqueous layer is removed to a new tube containing 10 μg of yeast tRNA and ethanol precipitated at −70°C for 30 minutes. Following centrifugation for 15 minutes the ethanol is discarded, the pellet is washed with 80% ethanol, dried, and resuspended in 6 μl of denaturing loading dyes. The samples are denatured for 5 minutes at 95°C, quick chilled on ice for 5 minutes and loaded on pre-electrophoresed (2000 V, 24 mA) 6–8% acrylamide:7 M urea sequencing type gels. When the bromophenol blue is close to the bottom of the gel, electrophoresis is stopped, the gel plates are separated, the gel wrapped in Saran wrap and autoradiographed at −90°C with an intensifying screen.

Primer extension[61]

cDNA synthesis with reverse transcriptase on mRNA is a modification of Treisman et al.[61] 20 μg of total RNA is annealed for 20 hours at 35°C to a uniformly labeled single stranded probe. The hybrids are ethanol precipitated, dissolved in 25 μl of 100 mM NaCl, 1 mM EDTA, 10 mM Tris, pH 8.2, and vortexed to resuspend the pellet. 25 μl of 10 mM DTT, 16 mM $MgCl_2$, 1 mM each deoxynucleotide triphosphate, 750 units/ml of Rnasin, and 100 mM Tris, pH 8.2, and 25 units of AMV reverse transcriptase (Seikagaiku) are added and incubated for 60 minutes at 40°C. EDTA is added to 10 mM and the RNA template is hydrolyzed with a final concentration of 1 N NaOH and incubation for 20 minutes at 70°C. This mixture is cooled on ice and neutralized with a final concentration of 1 N HCl, phenol extracted and ethanol precipitated. Pellets are resuspended in denaturing loading dyes, denatured at 95°C for 5 minutes, quick chilled, and the primer extension products are electrophoresed on 5–8% acrylamide:7 M urea sequencing gels.

REFERENCES

1 Chao M V, Mellon P, Charnay P, Maniatis T, Axel R. The regulated expression of β-globin genes introduced into mouse erythroleukemia cells. Cell 1983; 32: 483–493.
2 Charnay P, Mellon P, Maniatis T. Linker scanning mutagenesis of the 5′-flanking region of the mouse β-major-globin gene: sequence requirements for transcription in erythroid and non-erythroid cells. Mol Cell Biol 1985; 5(6): 1498–1511.
3 Dierks P, van Ooyen A, Cochran M D, Dobkin C, Reiser J, Weissmann C. Three regions upstream from the cap site are required for efficient and accurate transcription of the rabbit β-globin gene in Mouse 3T6 cells. Cell 1983; 32: 695–706.
4 Grosveld G C, Rosenthal A, Flavell R A. Sequence requirements for the transcription of the rabbit β-globin gene *in vivo*: the −80 region. Nucleic Acids Res 1982; 10(16): 4951–4971.
5 McKnight S L, Kingsbury R. Transcriptional control signals of a eukaryotic protein coding gene. Science 1982; 217: 316–324.
6 Allan M, Montague P, Gridlay G J, Sibbet G, Donovan-Peluso M, Bank A, Paul J. Tissue and stage specific transcription of the human ξ-globin gene following transfection into the embryonic erythroid cell line K562. Nucleic Acids Res 1985; 13: 6125–6136.
7 Banerji J, Olson L, Schaffner W. A lymphocyte-specific cellular enhancer is located downstream of the joining region in immunoglobulin heavy chain genes. Cell 1983; 33: 729–740.
8 Bank A, Donovan-Peluso M, Young K, Kosche K, Cubbon R, Dobkin C. Regulation of human globin gene expression. In: Stamatoyannopoulos G, Nienhuis A W, eds. Experimental approaches for the study of hemoglobin switching. New York: Alan R Liss, 1985.
9 Deschatrette J, Fougere-Deschatrette C, Carcos L, Schimke R T. Expression of the mouse serum albumin gene introduced into differentiated and dedifferentiated rat hepatoma cells. Proc Natl Acad Sci USA 1985; 82: 765–769.

10 Donovan-Peluso M, Acuto S, Swanson M, Dobkin C, Bank A. Expression of human γ globin genes in human erythroleukemia (K562) cells. Mol Cell Biol 1986; submitted.
11 Gillies S D, Morrison S L, Oi V T, Tonegawa S. A tissue-specific transcription enhancer element is located in the major intron of a rearranged immunoglobulin heavy chain gene. Cell 1983; 33: 717–728.
12 Kondoh H, Yasuda K, Okada T S. Tissue-specific expression of a cloned chick δ-crystallin gene in mouse cells. Nature 1983; 301: 440–442.
13 Ohlsson H, Edlund T. Sequence-specific interactions of nuclear factors with the insulin gene enhancer. Cell 1986; 45: 35–44.
14 Oi V T, Morrison S L, Herzenberg L A, Berg P. Immunoglobulin gene expression in transformed lymphoid cells. Proc Natl Acad Sci USA 1983; 80: 825–829.
15 Young K, Donovan-Peluso M, Bloom K, Allan M, Paul J, Bank A. Stable transfer and expression of exogenous human globin genes in human erythroleukemia (K562) cells. Proc Natl Acad Sci USA 1984; 81: 5315–5319.
16 Young K, Donovan-Peluso M, Cubbon R, Bank A. Trans acting regulation of β globin gene expression in erythroleukemia (K562) cells. Nucleic Acids Res 1985; 13(14): 5203–5213.
17 Church G W, Ephrussi A, Gilbert W, Tonegawa S. Cell-type-specific contacts to immunoglobulin enhancers in nuclei. Nature 1985; 313: 798–801.
18 Dynan W S, Tjian R. Control of eukaryotic messenger RNA synthesis by sequence-specific DNA-binding proteins. Nature 1985; 316: 774–778.
19 Dynan W S, Sazer S, Tjian R, Schimke R T. Transcription factor Sp1 recognizes a DNA sequence in the mouse dihydrofolate reductase promoter. Nature 1986; 319: 154–158.
20 Singh H, Sen R, Baltimore D, Sharp P A. A nuclear factor that binds to a conserved motif in transcriptional control elements of immunoglobulin genes. Nature 1986, 319: 154–158.
21 Graham F L, Abrahams P J, Mulder C et al. Studies on in vitro transformation by DNA and DNA fragments of human adenoviruses and simian virus 40. Cold Spring Harbor Symp Quant Biol 1975; 39: 637–650.
22 Mantei N, Boll W, Weissmann C. Rabbit β-globin mRNA production in mouse L cells transformed with cloned rabbit β-globin chromosomal DNA. Nature 1979; 281: 40–46.
23 Parker B A, Stark G R. Regulation of simian virus 40 transcription: sensitive analysis of the RNA species present early in infections by virus or viral DNA. J Virol 1979; 31(2): 360–369.
24 Wigler M, Silverstein S, Lee L S, Pellicer A, Cheng Y C, Axel R. Transfer of purified herpes virus thymidine kinase gene to cultured mouse cells. Cell 1977; 11: 223–232.
25 Schaffner W. Direct transfer of cloned genes from bacteria to mammalian cells. Proc Natl Acad Sci USA 1980; 77(4): 2163–2167.
26 Pearson T N, Pinder M, Roleants Kar S K, Lunsin L B, Mayor-Withey K S, Hewett R S. Methods for derivation and detection of antiparasite monoclonal antibodies. J Immunol Methods 1980; 34: 141–154.
27 Neumann E, Schaefer-Ridder M, Wang Y, Hofschneider P H. Gene transfer into mouse lyoma cells by electroporation in high electric fields. EMBO 1982; 1(7): 841–845.
28 Potter H, Weir L, Leder P. Enhancer-dependent expression of human κ immunoglobulin genes introduced into mouse pre-B lymphocytes by electroporation. Proc Natl Acad Sci USA 1984; 81: 7161–7165.
29 Zimmermann U, Vienken J. Electric field-induced cell-to-cell fusion. J Membr Biol 1982; 67: 165–182.
30 Dick J E, Magli M C, Huszar D, Phillips R A, Bernstein A. Introduction of a selectable gene into primitive stem cells capable of long term reconstitution of the hematopoietic system of W/W^v mice. Cell 1985; 42: 71–79.
31 Hock R A, Miller A D. Retrovirus-mediated transfer and expression of drug resistance genes in human haematopoietic progenitor cells. Nature 1986; 320: 275–277.
32 Joyner A, Keller G, Phillips R A, Bernstein A. Retrovirus transfer of a bacterial gene into mouse haematopoietic progenitor cells. Nature 1983; 305: 556–558.
33 Keller G, Paige C, Gilboa E, Wagner E F. Expression of a foreign gene in myeloid and lymphoid cells derived from multipotent haematopoietic precursors. Nature 1985; 318: 149–154.
34 Williams D A, Lemischka I R, Nathan D G, Mulligan R C. Introduction of a new genetic material into pluripotent haematopoietic stem cells of the mouse. Nature 1984; 310: 476–480.
35 Anderson W E. Prospects for human gene therapy. Science 1984; 226: 401–409.
36 Bank A, Donovan-Peluso M, LaFlamme S, Rund D, Lerner N. Approaches to gene therapy

for β thalassemia. In: Fucharoen S et al, eds. International Conference on Thalassemia. New York: March of Dimes Birth Defects Foundation, 1985.
37. Wigler M, Sweet R, Sim G K et al. Transformation of mammalian cells with genes from prokaryotes and eukaryotes. Cell 1979; 16: 777–785.
38. Wold B, Wigler M, Lacy E, Maniatis T, Silverstein S, Axel R. Introduction and expression of a rabbit β globin gene in mouse fibroblasts. Proc Natl Acad Sci USA 1979; 76(11): 5684–5688.
39. Pellicer A, Robins D, Wold B et al. Altering genotype and phenotype by DNA-mediated gene transfer. Science 1980; 209: 1414–1422.
40. Wigler M, Pellicer A, Silverstein S, Axel R, Urlab G, Chasin L. DNA-mediated transfer of the adenine phosphoribosyltransferase locus into mammalian cells. Proc Natl Acad Sci USA 1979; 76(3): 1373–1376.
41. Mulligan R C, Berg P. Expression of a bacterial gene in mammalian cells. Science 1980; 209: 1422–1427.
42. Mulligan R C, Berg P. Selection for animal cells that express the *Escherichia coli* gene coding for xanthine-guanine phosphoribosyl-transferase. Proc Natl Acad Sci USA 1981; 79(4): 2072–2076.
43. Pratt D, Subramani S. Nucleotide sequence of the *Escherichia coli* xanthine-guanine phosphoribosyl transferase gene. Nucleic Acids Res 1983; 11(24): 8817–8823.
44. Colbere-Garapin F, Horodniceanu F, Kourilsky P, Garapin A C. A new dominant hybrid selective marker for higher eukaryotic cells. J Mol Biol 1981; 150: 1–14.
45. Southern P J, Berg P. Transformation of mammalian cells to antibiotic resistance with a bacterial gene under control of the SV-40 early region promoter. J Mol Appl Genet 1982; 1(4): 327–341.
46. Blochlinger K, Diggelmann H. Hygromycin B phosphotransferase as a selectable marker for DNA transfer experiments with higher eucaryotic cells. Mol Cell Biol 1984; 4(12): 2929–2931.
47. Haber D A, Beverly S M, Kiely M L, Schimke R T. Properties of an altered dihydrofolate reductase encoded by amplified genes in cultured mouse fibroblasts. J Biol Chem 1981; 256(18): 9501–9510.
48. Kaufman R J, Sharp P A. Construction of a modular dihydrofolate reductase cDNA gene: analysis of signals utilized for efficient expression. Mol Cell Biol 1982; 2: 1304–1319.
49. Kaufman R J, Sharp P A. Amplification and expression of sequences cotransfected with a modular dihydrofolate reductase complementary DNA gene. J Mol Biol 1982; 159: 601–621.
50. Schimke R T, Kaufman R J, Nunberg J H, Dana S L. Studies on the amplification of dihydrofolate reductase genes in methotrexate-resistant cultured mouse cells. Cold Spring Harbor Symp Quant Biol 1979; 93: 1294–1303.
51. Simonsen C C, Levinson A D. Isolation and expression of an altered mouse dihydrofolate reductase cDNA. Proc Natl Acad Sci USA 1983; 80: 2495–2499.
52. Glisin V, Crkvenjakòv R, Byus C. Ribonucleic acid isolation by cesium chloride centrifugation. Biochemistry 1974; 13: 2633–2637.
53. Southern E M. Detection of specific sequences among DNA fragments separated by gel electrophoresis. J Mol Biol 1975; 98: 503–517.
54. Rigby P W J, Dieckmann M, Rhoades C, Berg P. Labeling deoxyribonucleic acid to high specific activity *in vitro* with DNA polymerase I. J Mol Biol 1977; 113: 237–251.
55. Goldberg D A. Isolation and partial characterization of the *Drosophila* alcohol dehydrogenase gene. Proc Natl Acad Sci USA 1980; 77(10): 5794–5798.
56. Thomas P S. Hybridization of denatured RNA and small DNA fragments transferred to nitrocellulose. Proc Natl Acad Sci USA 1980; 77(9): 5201–5205.
57. Melton D A, Krieg P A, Rebagliati M R, Maniatis T, Zinn K, Green M R. Efficient *in vitro* synthesis of biologically active RNA and RNA hybridization probes from plasmids containing a bacteriophage SP6 promoter. Nucleic Acids Res 1984; 12(18): 7035–7056.
58. Messing J. New M13 vectors for cloning. In: Wu R, Grossman L, Moldave K, eds. Methods in enzymology 101. New York: Academic Press, 1983: p 20–78.
59. Berk A J, Sharp P. Sizing and mapping of early adenovirus mRNAs by gel electrophoresis of S1 endonuclease digested hybrids. Cell 1977; 12: 721–732.
60. Weaver R F, Weissman C. Mapping of RNA by a modification of the Berk-Sharp procedure-5' termini of the 15S beta globin messenger RNA precursor and mature 10S beta globin messenger RNA have identical map coordinates. Nucleic Acids Res 1979; 6: 1175–1193.
61. Treisman R, Proudfoot N, Shander M, Maniatis T. A single-base change at a splice site in a β°-thalassemic gene causes abnormal RNA splicing. Cell 1982; 29: 903–911.

Glossary

Alkaline lysis procedure — a method for rapid isolation of bacterial plasmids and M13 replicative forms (RF). Magnesium ions in highly alkaline solutions will lyse cells and denature and precipitate linear (genomic) DNA molecules, but leave supercoiled circular plasmid and replicative form DNA in solution.

Antisense RNA — an RNA molecule designed to be complementary to a portion of an mRNA molecule. By annealing to the mRNA, the antisense RNA inhibits translation, thus blocking production of the protein product of the gene and its mRNA.

'Antisense' strand of DNA (see also 'sense' strand) — the strand of DNA that is complementary in base sequence to the strand (sense strand) that encodes the mRNA. It is thus usually a 'non-coding' strand in eukaryotes. Also called the '−', or 'Crick' strand.

Bacteriophage — a virus that infects a specific range of bacterial strains (e.g. λ bacteriophage infects certain strains of *E. coli*). The 'phage' (viral) particle is a very simple organism containing a DNA genome surrounded by 'coat' and 'tail' proteins. The DNA genome can be engineered and used as a gene cloning vehicle, since phage DNA replicates independently (episomal or extra-chromosomal DNA) during the typical lytic cycle (by which the virus lyses the host cell).

Bal 31 — an exonuclease whose rate of degradation of a DNA strand in a 3' to 5' direction can be controlled, making the enzyme useful for generating a population of DNA molecules from which graded portions of the 3 ' end have been removed.

Blot, blotting — a technique by which DNA, RNA or proteins are first separated into components by gel electrophoresis. The contents of the gel are then 'transferred' ('blotted') onto a membrane filter more suitable for hybridization analysis.

'CAP' — the 'CAP' is a modified oligonucleotide added to the 5' end of nascent mRNA molecules during processing in the nucleus. The 'CAP' enhances cytoplasmic stability and translation.

cDNA (copy DNA or complementary DNA) — a single stranded synthetic DNA molecule that is complementary in base sequence to the mRNA molecule from which it was copied (transcribed) by the enzyme 'reverse transcriptase' (RNA dependent DNA polymerase).

Chromatin — the complex in the vertebrate nucleus that contains the DNA and various proteins (histones and non-histone proteins) that interact with the DNA to form the highly compact structure of the chromosome.

'Cis-acting' element — a short DNA sequence (usually 10–50 bases long) that is the site of interaction between a 'trans-acting' factor and a gene. As a result, the level of expression of the gene is influenced. Promoter, enhancer, and silencer sequences are cis-acting elements.

Codon — a 3 nucleotide segment of mRNA that encodes an amino acid. During mRNA translation, ribosomes 'read' successive codons in the $5' \rightarrow 3''$ direction. The codons are recognized by a complementary three base sequence ('anticodon') positioned with a tRNA molecule that in turn is specifically bound to ('charged' with) the amino acid encoded by the codons.

Codon degeneracy — a term describing the fact that some amino acids can be encoded by two or more codons, e.g. 5'-UUU-3' and 5'-UUC-3' both encode phenylalanine. There is, however, no codon ambiguity; that is, one codon never encodes more than one amino acid.

Colony hybridization — a molecular technique in which bacterial colonies containing cloned species are lysed *in situ* on a solid support and DNA hybridized to a molecular probe which thereby locates specific colonies containing nucleotide sequences of interest. Specific colonies with these 'cloned' sequences can then be grown from replicas for further study and manipulation.

Denhardt's solution — 1 × Denhardt's = 0.02% (w/v) Ficoll, 0.02% (w/v) bovine serum albumin, 0.02% (w/v) polyvinylpyrrolidone. Used to block non-specific adherence of nucleic acid probes to blotting media.

Dideoxy nucleotides — also called ddNDPs or 'dideoxies', these deoxy nucleotide analogues have H-(deoxy) rather than -OH (oxy) groups attached to both the 2' and 3' positions of the sugar moiety of the nucleotide. They can be added to a growing strand via the 5' PO_4 group, and stop (terminate) DNA chain synthesis at the point at which they are incorporated because they have no 3' hydroxyl group to form the next phosphodiester bond. Used extensively for DNA sequencing and for 3' end labeling.

Dissociation temperature (T_d) *or melting temperature* (T_m) — the temperature at which 50% of the double stranded DNA, RNA or DNA : RNA molecules in a mixture are thermally denatured into single stranded form.

DEPC (diethyl pyrocarbonate) — non-specific RNase inhibitor.

DNA methylation — covalent modification of one of the four nitrogenous bases in the DNA polymer by the addition of a methyl group to one of the carbons of the nitrogenous base ring structure. In vertebrates, the methyl group is added to the carbon-5 position of the cytosine ring.

DNAse I (deoxyribonuclease I) — an endonuclease that is derived from bovine pancreas and that catalyzes the hydrolysis of the phosphodiester backbone of DNA molecules.

DNA or RNA probe — a term used to refer to DNA or RNA molecules tagged with readily detected labels (e.g. radioisotopes or biotin). The tagged molecules will hybridize only to those DNA or RNA species in a sample having complementary base sequences. The labeled molecules thus serve to mark or identify only the complementary molecules.

3′ end — the end of a DNA or RNA strand that has a free 3′ hydroxyl or phosphate residue on the sugar moiety of the end nucleotide of the strand. All other nucleotides, except the 5′ end nucleotide, have the 5′ and 3′ position bound in phosphodiester bonds.

5′ end — the end of a DNA or RNA strand that has a free 5′ hydroxyl or phosphate residue on the sugar moiety of the end nucleotide of the strand. All other nucleotides, except the 3′ end nucleotide, have the 3′ and 5′ position bound in phosphodiester bonds.

Endonuclease — a nuclease that cuts (cleaves) a DNA or RNA molecule at interior positions along the chain. Different from exonucleases, which cut or 'nibble' DNA or RNA molecules, processively only from one end.

Enhancer — a short DNA sequence — located near or, sometimes, in a gene — that causes increased activity (presumably by interaction with appropriate stimulatory nuclear proteins called 'trans-acting' factors) of the gene in the appropriate tissues.

Ethidium bromide (EtBr) — a pinkish red compound that intercalates into the helices of DNA and RNA strands. When intercalated in this manner, EtBr fluoresces a bright pink-orange color upon exposure to ultraviolet light. Used to stain DNA and RNA bands in gels, and to quantitate nucleic acid samples.

Exonucleases — enzymes that shorten or 'nibble' DNA molecules by removing bases successively (processive cleavage) from one end of the molecule toward the other. 5′ exonucleases attack the 5′ end and 'nibble' toward the 3′ end. 3′ exonucleases attack from the 3′ end toward the 5′ end.

Formamide — a reversible, relatively innocuous (to DNA and RNA) denaturing agent that destabilizes (lowers the T_m of) double stranded hybrids. Useful for performing hybridization and 'melting' of nucleic acid strands at lower temperatures than achievable in aqueous buffers alone.

λgt11 — λ bacteriophage engineered so that it can produce proteins encoded by recombinant DNA sequences inserted into the phage genome. A representative example of an 'expression cloning vector', λgt11 is widely used to prepare cDNA libraries for screening by antibodies against the probe encoded by the gene one wishes to isolate. A relative, λgt10 cannot express recombinant proteins but is a more efficient cloning vector.

G-C content, A-T content — the percentage of the bases in a DNA molecule that form G-C or A-T base pairs, respectively. G-C 'rich' molecules are more thermally stable (higher T_m).

Genomic DNA — DNA that has been extracted directly from the nucleus of a particular cell type or tissue of origin. This is the DNA that forms the chromosomes of the nucleus, and is distinct from cDNA, which is a synthetic DNA transcribed from an mRNA template, or episomal DNA (e.g. plasmids, certain viruses) which exist in cells independently of the bulk of the genome.

Guanidine hydrochloride — chaotropic agent used to inhibit RNase, not as powerful as guanidinium thiocyanate.

Guanidinium thiocyanate — powerful chaotropic agent, used to inactivate RNase.

Hybridization — this term refers to the binding of one DNA or RNA strand to a strand with the complementary base sequence to form a double stranded hybrid. The term 'annealing' is often used interchangeably with hybridization. The two strands of RNA and DNA are held together by hydrogen bonds. Stable hydrogen bonds form between strands only according to the rules of Watson-Crick base pairing (G=C or A=T bonds form, but A-C, A-G, C-T, G-T, A-A, C-C, G-G and T-T bonds do not).

Hybridization selection — a molecular technique in which RNA species are selectively hybridized to a cloned DNA species, effecting purification of a complementary RNA for subsequent *in vitro* translation.

In situ hybridization — detection of mRNA (in cells or tissues) or DNA (in chromosome spreads) sequences by hybridization of a probe directly to a tissue cell or chromosome preparation on a microscope slide. Allows one to visualize the cell-to-cell or point-to-point localization and distribution of the sequence being studied.

In vitro packaging reaction — a biochemical mixture of *E. coli* extracts capable of assembling phage protein components around a recombinant bacteriophage DNA molecule, thus 'packaging' the DNA into a viable, infectious bacteriophage particle.

IPTG (isopropylthiogalactoside) — a potent inducer of *E. coli* β-galactoside gene.

'Klenow' enzyme or fragment — a portion (fragment) of *E. coli* DNA polymerase that contains the elongating activity useful for polymerizing DNA strands from single stranded DNA templates using a complementary primer oligonucleotide.

Lysogen — a bacterial cell carrying a λ bacteriophage genome which has integrated into its genome and thus inactivated its lytic functions. The bacteria thus survives infection. Yet it can express many genes in the phage genome, including genes containing cloned cDNA fragments.

mRNA, messenger RNA — the RNA species which codes for the translation of proteins in ribosomes; mRNA is transcribed by RNA polymerase II from DNA in the nucleus, then processed by splicing out of introns, addition of the poly-A tail, and capping (addition of a 'CAP' oligonucleotide to the 5' end).

M13 bacteriophage — a very useful bacterial virus for generating single stranded DNA molecules containing cloned DNA segments. The virus exists as a double stranded DNA molecule in the *E. coli* host cell, but excretes phage particles containing only 1 of the 2 strands (the '+' strand) into the medium. Used for DNA sequencing, *in vitro* mutagenesis and preparation of hybridization probes.

Melting temperature (T_m) or dissociation temperature (T_d) — the temperature at which 50% of the double stranded DNA, RNA or DNA : RNA molecules in a mixture are thermally denatured into single stranded form.

Nick translation — a method for incorporating labeled nucleotides into a double stranded DNA molecule. The molecule is first nicked by exposure to small amounts of DNase that make random single stranded cuts (single stranded nicks) at numerous points in the molecule. *E. coli* DNA polymerase I and radioactivate dNTPs are then added. As it repairs the nicks, the polymerase incorporates the labeled dNTPs into the DNA.

Northern blotting, Northern blots — procedure for analyzing RNA transcripts, analogous to Southern blotting; the RNA to be analyzed is fractionated according to size (molecular weight) on agarose gels, then blotted onto a nitrocellulose or nylon membrane. The blot is then probed for the presence of a specific RNA species by incubation with a labeled DNA or RNA probe complementary in base sequence to that RNA species.

Nuclease — the general term for an enzyme which degrades mRNA and/or DNA.

NZYM buffer — Nzyamine medium and yeast tryptone extract (available commercially, see N-Z medium) plus 0.4% maltose.

Oligo-dT cellulose — a cellulose resin onto which an oligo-dT polymer (short DNA molecules 12–20 residues long, containing only Ts) is attached. Used to purify mRNA because most mRNAs contain long 'tails' of 'poly-A' at their 3' ends.

Oligomer — an oligonucleotide.

Oligonucleotide — a small DNA molecule, usually ranging from a few bases to about 150 bases long. The term is usually used to refer to single stranded DNA segments assembled by DNA synthesizing machines. Examples include synthetic oligonucleotides designed to be complementary to specific regions of genes for use as primers, for sequencing of polymerase reactions, and synthetic 'linkers' containing useful restriction endonuclease sites.

Oligonucleotide-directed site-specific mutagenesis — a procedure by which any cloned DNA sequence can be altered at specific sites by use of a short fragment of DNA (the oligonucleotide) containing the desired base substitutions.

Plasmid — a circular DNA molecule with its own origin of replication, capable of co-existing and replicating independently of the genomic DNA in the cell (extra-chromosomal DNA element, episome). In microbes, plasmids useful for molecular biology carry antibiotic resistance markers, convenient restriction endonuclease sites, and other useful DNA sequence regions such as translation start sites, etc.

Poly-A^+ RNA — RNA molecules having sufficiently long stretches of polyadenylate (A) residues to bind to oligo-dT cellulose columns. The major (but not only) poly-A^+ RNAs in cells are mRNA molecules. Poly-A^+ RNA is often the fraction used as an 'mRNA' preparation.

Poly-A^+ selection — procedure used to enrich an RNA preparation for mature mRNA by utilizing a stretch of deoxythymine residues attached to a solid support (usually cellulose).

Poly-A^+ tail — a stretch of adenosine residues attached to the 3' end of the mature mRNA and not present in other RNA species.

Polylinker — an oligonucleotides 10–50 bases long designed to include a large number of restriction endonuclease sites; when inserted in a plasmid, a linker creates convenient cloning sites.

Polymerase chain reaction — an enzymatic method for selectively amplifying one specific DNA sequence within a complex mixture. One chooses the sequence (at this time, up to about 5 kb long) to amplify, and constructs oligonucleotides complementary to short sequences (15–20 bases long) flanking each end of the desired sequences. The primers are annealed to the DNA, and DNA polymerase is then added. At the end of the reaction the DNA is denatured again and the annealing and polymerase steps repeated. Since synthesis of the daughter strands can proceed only between the primers, only the region of DNA between the primers is amplified.

Primer extension — a method for synthesizing a cDNA molecule complementary only to a single mRNA or region of an mRNA or DNA molecule by using an oligonucleotide complementary to a defined mRNA region as template for the polymerase or reverse transcriptase reaction.

Promoter — the region in the 5' ('upstream') flanking region of gene that serves as the region recognized by RNA polymerase II and used as the entry point of the enzyme for establishment of a transcription complex. Most eukaryotic promoters contain the sequence 'TATA' about 20–30 bases from the actual point at which transcription starts ('CAP site'). Many promoters contain the sequence CCAAT at about 80 bases upstream and additional regions 100–200 bases upstream rich in G-C base pairs.

Proteinase K — a powerful proteolytic enzyme that is isolated from *Aspergillus oryzae*. It is active in the presence of SDS. Useful for removing proteins (especially nucleases) from mixtures containing nucleic acids; residual degraded protein and proteinase K can be removed by phenol : chloroform extraction.

PSB buffer — phosphate buffered saline:
    0.14 M NaCl
    0.01 M NaPO$_4$, pH 7.0
pH balanced and isotonic buffer used for washing cell and biochemical preparations.

Random primer labeling method — double stranded DNAs are denatured to single stranded form and annealed to a mixture of all possible combinations of hexamers (6 base oligonucleotides) containing the four bases dA, dC, dG, dT. The Klenow fragment of DNA polymerase I and labeled dNTPs are added. The polymerase uses the random hexamers as primers and the denatured strands as templates to synthesize labeled daughter strands statistically. There are many random sites on almost any DNA template strand that can bind with enough hexamers in the mixture to prime efficient polymerization.

Replica plating — the precise, templated transfer of bacterial or bacteriophage colonies or plaques from one plate or solid support to another.

Restriction endonucleases — bacterial enzymes which recognize and cut specific sequences in double stranded DNA. These enzymes cleave DNA only at positions bearing defined short base sequences, usually 4, 5, or 6 bases long (e.g. Eco RI cuts DNA only at sites where the sequence 5'-GAATTC-3' is present).

RNA or DNA probe — a term used to refer to DNA or RNA molecules tagged with readily detected labels (e.g. radioisotopes or biotin). The tagged molecules will hybridize only to those DNA or RNA species in a sample having complementary base sequences. The labeled molecules thus serve to mark or identify only the complementary molecules.

RNase(s) — enzymes which degrade RNA; generally extremely hardy and difficult enzymes to inactivate.

RNase protection assay — procedure for analyzing RNA transcripts, in which the RNA is hybridized to a single stranded radioactively labeled RNA probe; RNase is then used to digest any portions of the probe which are not protected by being bound within a double stranded helix. The products of the reaction are electrophoresed in an acrylamide gel for analysis.

mRNA, messenger RNA — the RNA species which codes for the translation of proteins in ribosomes; mRNA which is transcribed by RNA polymerase II from DNA in the nucleus, then processed by splicing out of introns, addition of the poly-A tail, and capping (addition of a 'CAP' oligonucleotide to the 5' end).

SDS (sodium dodecyl sulfate) — a potent ionic detergent.

'Sense' strand of DNA — the strand of DNA that is complementary in base sequence to its mRNA transcript, i.e. the strand that codes for, or serves as template for the mRNA. Also called the coding '+' or 'Watson' (as opposed to the (−), non-coding, or 'Crick' strand).

SET buffer — 1 × SET = 0.15 M NaCl
$\qquad$ 1 mM EDTA
$\qquad$ 0.03 M Tris-HCl, pH 8.0

SET is a standard buffer used for DNA and RNA hybridization blotting.

Silencer — a short DNA sequence that tends to cause repressed DNA activity if bound by appropriate proteins. (See *enhancer*.)

SM buffer — 10 mM MgSO$_4$
$\qquad$ 100 mM NaCl
$\qquad$ 50 mM Tris HCl, pH 7.5
$\qquad$ 0.02% gelatin

This medium is used for the dilution and storage of bacteriophage λ.

Southern blotting — a technique which allows one to detect and characterize specific DNA sequences. DNA is digested with a restriction enzyme; the resulting fragments are separated by size on an agarose gel. The DNA fragments are transferred to nitrocellulose filter paper, and the nitrocellulose filter is hybridized to a radiolabeled fragment containing the DNA sequence of interest.

Southwestern blot — Protein is first blotted onto a membrane filter (Western blot). The filter is then incubated with radioactively labeled DNA in order to determine which protein bands bind in a specific manner to DNA sequences.

SP6 and T$_7$ RNA polymerases — enzymes isolated from phage-infected bacteria that synthesize RNA from DNA templates only at sites downstream from specific promoter sequences.

SSC — 1 × SSC = 0.15 M NaCl
$\qquad$ 0.015 M sodium citrate, pH 7.0

The standard salt and buffer system used for blotting and annealing reactions.

Staphylococcal protein A — a protein that binds to antigen antibody complexes since it can be readily purified and labeled, e.g. with ^{125}I, the protein is used to locate those proteins that have complexed to an antibody probe.

Subtraction hybridization — DNA or RNA in single stranded form from two different sources are allowed to hybridize; complementary double stranded species common to nucleic acids from each source are then separated from the single stranded molecules which contain sequences unique to each source.

T$_4$ DNA polymerase — an enzyme that acts as a $3' \rightarrow 5'$ exonuclease if exposed to DNA in the absence of deoxynucleotide triphosphates; if dNTPs are added back before the DNA molecule is completely digested, the enzyme becomes a polymerase, using the single stranded regions exposed by the prior exonuclease action as template. T$_4$ polymerase provides a very handy way to label DNA by incorporating high specific activity ^{32}p or ^{35}S dNTPs.

T$_4$ polynucleotide kinase — an enzyme that adds a phosphate group to the 5'-OH residue at the 5' end of a DNA molecule and provides an excellent way to 'end-label' the 5' extremity of a DNA molecule or nucleotide.

T$_7$ polymerase — a very useful enzyme that synthesizes DNA strands from single stranded templates annealed to a short primer oligonucleotide. A commercial brand of the enzyme called 'sequenase' is becoming the standard reagent for dideoxy DNA sequencing.

TE buffer — 10 mM Tris, ph 7.5
    1 mM EDTA
Commonly used for dissolving and storing DNA and RNA solutions.

TEA buffer — 40 mM Tris acetate
    2 mM EDTA
Commonly used for DNA gel electrophoresis.

TEB buffer — 0.089 M Tris borate
    0.01 M EDT. pH 7.5
    0.089 M boric acid
Commonly used for DNA gel electrophoresis.

Terminal transferase (TdT) — an enzyme that adds bases to the 3' end of double or single stranded nucleic acids without a need for a primer. Used to add 3' single stranded 'tails', it can also be used to 3' end label DNA strands by restricting the length of tails that can be added to one or a few bases with dideoxy NTPs or cordycepin.

Tetramethylammonium chloride — a chemical whose inclusion in high concentration in nucleic acid hybridization reactions ablates the tendency of DNA sequences with higher percentages of guanine (G) and cytosine (C) pairs to bind more avidly.

Top agar — a warm liquid agar or agarose solution in which bacteria and bacteriophage are suspended. The suspension is poured over a nutrient agar substratum and allowed to solidify.

Total cellular RNA — a sample including all RNA species from a cell, including messenger RNA (mRNA), ribosomal RNA (rRNA), transfer RNA (tRNA), and other minor species of RNA.

'Trans-acting' factor — a term applying to a presumed molecule or molecular complex that binds to one or more DNA sequence elements (e.g. promoter, enhancer, or silence-sequence called a 'cis-acting' element) and thereby influences the level of transcription of the gene or genes located near the DNA sequence element. Most trans-acting factors are assumed to be nuclear proteins or protein-nucleic acid complexes.

Translation — converting the genetic code contained in the primary sequence of the mRNA into a specific protein sequence. This can be accomplished in cells by ribosomes and tRNAs which form polyribosomes that read the mRNA code like a ticker tape. (See also *Codon*.)

Western blot — a blot that is used to analyze protein, using an antibody probe, after separation of proteins by gel electrophoresis. (See *Northern, Southern blotting*.)

X-Gal (5-bromo-4-chloro-3-indolyl-β-D-galactoside) — a chromogenic substrate of β-galactoside that turns blue when cleaved by the enzyme.

YT broth (yeast tryptone broth) — standard 'rich' bacterial growth medium.

Index

A-T content, G-C content, 161
Alkaline lysis procedure, 155
Aminopterin, 182
Analysis of transfected cells, 187
   DNA analysis, 187, 188
      Southern blotting and hybridization, 188
   harvesting cells, 187
   RNA analysis, 189
      RNA dot-blots, 189
      Northern blotting and hybridization, 189
      SP6 RNAse protection, 190
      S_1 analysis, 191
      primer extension analysis, 192
Antifoam-A, 41
'Antisense' strand of DNA, 159
8-azaadenine, 182

Bacteriophage, 93–97
   DNA isolation, 73
      minilysates, 73
   DNA genomic cloning, 81–87
   λ, 108
   λ gt 11, 100
   λ gt 11, λ gt 10 cloning, 79–81
   M13, 124, 149, 150
Bal 31, 132
Blot, blotting
   Northern, 165, 189
   Southern, 21, 30, 163, 188
   Southwestern, 201
   Western, 203
5' bromodeoxyuridine, 182
Buffers, 25
   Klenow, 77
   NZYM, 74
   PSB, 201
   RT, 76
   SET, 91
   SM, 93
   SSC, 45, 164–167
   TA, 80
   TE, 23, 184
   TEA, 114
   TEB, 31
   TNE, 106

Tris/EDTA/Sarkosyl, 39, 45
Urea buffer/LiCl (lithium chloride), 40–42
Buffy coat (of peripheral blood samples), 40

cDNA (copy DNA or complementary DNA), 75, 196
   cloning, 72
      λ gt 11, λ gt 10 bacteriophage, 79
      plasmids, 72
   synthesis, 75, 79
      second strand, 77, 79
   libraries, 75–81
Chain reaction, 61
   polymerase, 199
Chromatin, 111, 118
Cis-acting elements, 112, 181
Cloning
   cDNA plasmids, 72
   λ gt 11, λ gt 10 bacteriophage, 79
Codon, 160–161
Codon degeneracy, 159
Colony hybridization, 89–93
cRNA, 115

ddNTPs, 124
Denhardt's solution, 45, 115, 162
DEPC (diethyl pyrocarbonate), 38
2,6 diaminopurine, 182
Dideoxy method of sequencing, 124, 130
   DNA sequencing, 125
Dideoxy nucleotides, 124
Dissociation temperature (T_d), 58, 59, 162
DNA
   antisense strand, 159
   cDNA, 72
   generation of blunt ends, 80
   genomic, 81
      bacteriophage cloning, 81, 83
      cosmid cloning, 85
   ligation
      to λ phage, 81
      linker, 80
   methylation *in vitro*, 80
   'sense' strand, 20
   transformation, *E. coli*, 73

DNA analysis of transfected cells, 187, 188
  Southern blotting and hybridization, 188
DNA, isolation of, 21, 112
  bacteriophage, 73
    minilysates, 73
    plasmids, 75, 76
  detecting phenol contamination, 24
  detecting protein contamination, 24
  methylation, 112
  very high molecular weight >50 kb, 23
DNA polymerase
  Klenow fragment, 53, 124, 149, 150
  T_4, 52
DNA probe, 158
DNase I (deoxyribonuclease I)
  hypersensitive site, 111, 118, 120

E. coli strains as hosts for M13 phage, 127, 151
E. coli, DNA transformation, 79
End labeling, 66–68
  indirect, 120
3' end, 112, 120–122
5' end, 111, 120–122
Endonuclease, 112
  Restriction, 25, 112
    methylation sensitive, 120
    stock buffers, 28
Enhancer, 197
EtBr (ethidium bromide), 44
Exonuclease, 197
Expressing of recombinant proteins in E. coli, 124

Fibrinogen, 93, 96
Formamide, 197
  effect on melting temperature, 34

λ gt 11 bacteriophage, 101
λ gt 11, λ gt 10 cloning, 79–81
G-C content, A-T content, 161
Gel electrophoresis, 30–32
  RNA, formaldehyde/agarose, 48
Gene transfer, 181, 183
  calcium phosphate precipitation, 181, 183
  electroporation, 181, 186
  lymphocyte separation medium (1sm), 184
  protoplast fusion, 181, 185
  retroviruses, 187
  see also Selection strategies
Genomic DNA, 174
  bacteriophage cloning, 81, 83
  cosmid cloning, 82
Genomic sequencing, 116
Genticin (G418), 183
Globin mRNA, 177, 178
Guanidine hydrochloride, 39
Guanidinium thiocyanate, 39–40, 46

Hemin, 171
  used in in vitro translation, 171
High molecular weight DNA >50 kb, 22–25
  DNA >30 kb, 112
Heterogeneous nuclear RNA, 37
Hybrid-arrested translation, 174
Hybrid-selected translation, 176
Hybridization, 33–36, 158
  colony, 89–93
  in situ, 139
    tissue preparation, 140
  RNA probes, 62, 142
  selection, 98
  Southern blot protocol, 31
  stringency, 92
  subtraction, 98
Hydroxyapatite chromatography, 98
Hygromycin B, 181, 183
Hypomethylation, 111
Hypoxanthine, 182
Hypoxanthine phosphoribosyl transferase, 182

Immunoprecipitation, 98
Indirect end labeling, 120
In situ hybridization, 139
In vitro
  methylation, DNA, 111
  packaging reaction, 81
  translation, 169
    hemin, 171
    mRNA, 37
IPTG (isopropylthiogalactoside), 126
Isolation
  DNA, 21–25, 112
    bacteriophage, 73
      minilysates, 73
    ethanol precipitation, 22, 42, 48
    high molecular weight, 21–25
    plasmids, 73
  mRNA, 37, 173
  nuclei, 118

Klenow fragment of DNA polymerase, 53, 56, 124, 129, 150
  enzyme or fragment, 200

Labeling methods
  end labeling, 66–68
  indirect, 120
  random primer, 54, 115
Libraries of recombinant DNA molecules
  amplification, 88, 89
  cDNA, 75–81
  cosmid, 85–87
  genomic, 81–87
  phage, 83–85
LiCl (Lithium Chloride)/Urea buffer, 40–42
  RNA extraction, 47
Ligation
  DNA linker, 81

to λ phage, 81
T-lymphocyte antigen receptor, 98
Lysogen, 106
Lysozyme, 91

mRNA, messenger RNA, 37
  in vitro translation systems, 170
  isolation, 37, 173
  oligo dT chromatography of, 42, 49
  oocyte, 170, 171
  purification, 49
  rabbit reticulocyte lysate, 171
  wheat germ, 170
M13 bacteriophage, 124, 148
Melting temperature (T_m), 34, 162
Meta-cresol precipitation, 24
Methotrexate, 183
2-methoxyethanol extraction of DNA, 24
Methylation, DNA, 111
  in vitro, 80
Mutagenesis, site-specific, 148
Mycophenolic acid, 182

Nick translation, 52, 115
Northern blotting, Northern blots, 16, 38, 43–44, 52, 165
Nuclease, 172
Nuclei, isolation, 118
NZYM buffer, 74

Oligo-dT (oligo-deoxythymidylate), 42–43
  cellulose, 42
  chromatography of mRNA, 42
Oligomer, 42
Oligonucleotide, 51, 55, 97, 148, 158
  directed site-specific mutagenesis, 112
  probe , 56–57
  synthetic, 55

Phage, 74, 125
Phenol : chloroform extraction for DNA isolation, 22
  ethanol precipitation, DNA isolation, 23
Plaque, 126
Plasmid, 101
  cDNA cloning, 75
  DNA isolation, 72, 73
  DNA preparation, 64, 65
Poly-A, poly-A tail, 42, 43
  poly-A selection, 12, 38
Poly (A)$^+$RNA, 59, 164
Polyethylene glycol, 74
Polylinker, 199
Polymerase
  chain reaction, 60
  SP6 and T7 RNA, 62
  T_4 DNA, 52
  T_7, 62
Polynucleotide kinase, T_4, 66
Polyribosomes, 98

Polytron (Brinkmann, Westbury, New York), 39, 41
Primer construction, 159
Primer extension, 167, 192
Probes
  DNA, 51, 158
  RNA, 63, 158
Promoter, 111, 200
Proteinase K, 21, 112
PSB buffer, 200

Quantitation of DNA, 24–25
  by spectrophotometry, 24–25
  by ethidium bromide fluorescence, 24–25

Random primer labeling method, 54, 115
Replica plating, 90–91
Restriction endonuclease, 25, 112
  methylation sensitive, 114
Restriction enzymes, 25
  Type I, 25
  Type II, 25–26
  cohesive and blunt ends, 25
  'star' activity, 26
  optimal time and temperature conditions, 27, 28
  stock buffers for restriction endonucleases, 28
Restriction mapping, 25
Ribosomal RNA (rRNA), 37
RNA
  'antisense', 159
  hn RNA, 37
  hybridization probes, 62, 140
  mRNA (messenger RNA), 37
  Poly-(A)$^+$, 164
  probes, 158
  rRNA (ribosomal RNA), 37
  SP6 and T_7 polymerases, 62–64
  total cellular RNA, 37
RNA analysis of transfected cells, 189
  RNA dot-blots, 189
  Northern blotting and hybridization, 189
  SP6 RNase protection, 190–191
  S_1 analysis, 191
  primer extension analysis, 192
RNase, 37, 38
  protection assay, 63

SDS (sodium dodecyl sulfate), 21
Selectable genes, 181, 182
  dominant, 182
Selection strategies
  complementation, 182
    adenine phosphoribosyl transferase (APRT), 182
    thymidine kinase, 182
  Dominant selection, 182
    aminoglycoside 3′ phosphotransferase (NEO), 183

Selection strategies (contd)
    dihydrofolate reductase (DHFR), 183
    hygromycin B phosphotransferase (HPH), 183
    xanthine-guanine phosphoribosyl transferase (XGPRT), 182
Selection, Poly-A hybridization, 98
'Sense' strand of DNA, 201
Sequenase, 131
Sequencing, 124
    dideoxy method, 130
SET buffer, 91
SM buffer, 93
Southern blotting, 21, 112, 114, 163
    agarose gel electrophoresis, 30–32
    transfer method, 32
    hybridization protocol, 33–36
Southwestern blotting, 201
SP6 and T_7 RNA polymerases, 18, 62–64, 143
SSC buffer, 202
Staphylococcal Protein A, 100
Strands of DNA
    'antisense', 159
    'sense', 201
Stringency hybridization, 92
Subtraction hybridization, 98
Synthesis
    cDNA, 75–77, 79
    cDNA, second strand, 77, 79–80

T_4 DNA polymerase, 52–53
T_4 polynucleotide kinase, 201

T7 polymerase, 62, 63, 201
T7 and SP6 RNA polymerases, 62–64, 143, 201
TE buffer, 202
TEB buffer, 31
Terminal transferase (TdT), 78, 202
Tetramethylammonium chloride, 97
Tissue
    DNA extraction from, 21
    preparation for *in situ* hybridization, 140
    RNA extraction from, 37
Top agar, 90
Total cellular RNA, 37
Trans-acting factors, 112, 181
Transfer RNA (tRNA), 37
Translation, 98, 169
    Hybrid-arrested, 174
    Hybrid-selected, 175
    in vitro with hemin, 171
    nick, 53
Tris/EDTA/Sarkosyl buffer, 39, 45
    lysis of RNA, 45–46

Urea buffer/LiCl (lithium chloride), 40, 42, 47

Western blotting, Western blots, 203

X-Gal (5-bromo-4-chloro-3-indolyl-β-D-galactoside), 126

Yeast tryptone broth (YT broth), 136, 203